PLATO, NOT PROZAC!

FROM THE LIBRARY OF

PLATO
NOT PROZAC!

APPLYING PHILOSOPHY
TO EVERYDAY PROBLEMS

LOU MARINOFF, PH.D.

HarperCollins*Publishers*

HarperCollins books may be purchased for educational, business, or sales promotional
use. For information please write: Special Markets Department, HarperCollins
Publishers, Inc., 10 East 53rd Street, New York, NY 10022.

Designed by Nancy Field

Library of Congress Cataloging-in-Publication Data

Marinoff, Lou.
    Plato, not prozac! : applying philosophy to everyday problems / Lou Marinoff.
— 1st ed.
     p. cm.
    Includes bibliographical references and index.
    ISBN 0-06-019328-X
    1. Philosophical counseling.   I. Lou Marinoff. II. Title.
BJ1595.5M37   1999
   100—dc21   99-22650

99 00 01 02 03 ❖/RRD 10 9 8 7 6 5 4 3

To those who always knew philosophy was good for something,
but could never say exactly what

"It is Reason alone which makes life happy and pleasant, by expelling all false Conceptions or Opinions, as may in any way occasion perturbation of mind."
—EPICURUS

"The unexamined life is not worth living."
—SOCRATES

"The time of human life is but a point, and the substance is a flux, and its perceptions dull, and the composition of the body corruptible, and the soul a whirl, and fortune inscrutable, and fame a senseless thing. . . . What then is there which can guide a man? One thing and only one, philosophy."
—MARCUS AURELIUS

"*The clouds of my grief dissolved and I drank in the light. With my thoughts recollected I turned to examine the face of my physician. I turned my eyes and fixed my gaze upon her, and I saw that it was my nurse in whose house I had been cared for since my youth—Philosophy.*"

—ANICIUS BOETHIUS

"*Man cannot overestimate the greatness and power of his mind.*"

—GEORGE FREDERICK HEGEL

"*To do philosophy is to explore one's own temperament, yet at the same time to attempt to discover the truth.*"

—IRIS MURDOCH

"*Carpenters fashion wood; fletchers fashion arrows; the wise fashion themselves.*"

—BUDDHA

Contents

PART III
BEYOND CLIENT COUNSELING

PART IV
ADDITIONAL RESOURCES

Acknowledgments

Thanks to my philosophical predecessors and contemporaries for their perennial inspiration. Philosophy is an endless river, meandering here and flowing there, but nowhere running dry.

Thanks to many academic and professional colleagues, in America and abroad, for their constructive interchange. They sustain the flame of philosophical inquiry, via illuminating theories and efficacious practices.

Thanks to all practitioners who contributed case studies. Owing to the usual constraints, we could not embody every submission. We have used cases from Keith Burkum, Harriet Chamberlain, Richard Dance, Vaughana Feary, Stephen Hare, Alicia Juarrero, Chris McCullough, Ben Mijuskovic, Simon du Plock, and Peter Raabe. Thanks also to those practitioners whose work or insights I have had occasion to mention: Gerd Achenbach, Stanley Chan, Pierre Grimes, Kenneth Kipnis, Ran Lahav, Peter March, and Bernard Roy.

Thanks to our Dutch colleagues—especially Dries Boele and Ida Jongsma—for training the first nucleus of American facilitators in the Nelsonian method of Socratic Dialogue.

Thanks to many others whose clear vision and unstinting support have furthered the emergence of philosophical practice in America—including Charles DeCicco, Joëlle Delbourgo, Ruben Diaz Jr., Paul del Duca, Ron Goldfarb, John Greenwood, Robbie Hare, Mahin Hassibi, Merl Hoffman, Ann Lippel, Thomas Magnell, Robyn Leary Mancini, Jean Mechanic, Thomas Morales, Yolanda Moses, Gerard O'Sullivan, Mehul Shah, Paul Sharkey, Wayne Shelton, Jennifer Stark, Martin Tamny, and Emmanuel Tchividjian.

Finally, thanks to Colleen Kapklein, who skillfully translated my elliptical ramblings into accessible prose. Any philosopher, I have learned, can write an unpopular book completely unaided. After all, it is our gift to take simple matters and make them astonishingly complex. By contrast, popular writing requires the artful yet faithful rendering of complex matters into astonishingly simple ones—a task I could not have conceived, let alone executed, without expert assistance.

LOU MARINOFF
New York City, 1999

• • •

The case studies herein are drawn from my counseling practice and from the practices of colleagues who voluntarily submitted them for inclusion. The anonymity of our clients has been protected by altering names, places, occupations, details, and other pertinent information. Although their identities have been fictionalized, the philosophical benefits they received are real.

PART · I

NEW USES FOR
ANCIENT WISDOM

1

What Went Wrong With Philosophy—and What's Going Right With It Lately

*"As for Diseases of the Mind, against them Philosophy
is provided of Remedies; being, in that respect, justly
accounted the Medicine of the Mind."*
—EPICURUS

*"To be a philosopher is not merely to have subtle thoughts, nor
even to found a school. . . . It is to solve some of the problems of
life, not theoretically, but practically."*
—HENRY DAVID THOREAU

A young woman confronts her mother's terminal breast cancer. A man contemplates a midlife career change. A Protestant woman whose daughter is engaged to a Jewish man and whose son is married to a Muslim woman fears potential religious conflicts. A successful business executive struggles over whether to leave his wife of over twenty years. A woman is happily living with her partner, but only one of them wants children. An engineer and single father supporting four children is afraid that blowing the whistle on a design flaw in a high-pressure project could cost him his job. A woman who has everything she thought she wanted—loving husband and children, beautiful house, high-paying career—struggles with meaninglessness: when she looks at her life she thinks, "Is this all there is?"

All of these people have sought professional help in managing the problems they feel overwhelming them. In another day, they might have found their way to the offices of a psychologist, psychiatrist, social worker, marriage counselor, or even their general practitioner for help with "mental illness." Or they might have consulted a spiritual adviser or turned to religion for moral instruction and guidance. And some of them may have been helped in those places. They may also have endured discussions about their childhoods, analysis of their behavior patterns, prescriptions for antidepressants, or arguments about their sinful nature or God's forgiveness, none of which got to the heart of their struggle. And they may well have been signed up for a lengthy and open-ended course of treatment, with a focus on diagnosing the illness as though it were a tumor to be excised or a symptom to be controlled with drugs.

But now there's another option for people unsatisfied by or opposed to psychological or psychiatric therapy: philosophical counseling. What the people described above did was seek out a different kind of assistance. They consulted a philosophical practitioner, looking for insight from the world's great wisdom traditions. As established religious institutions lose their authority with more and more people, and as psychology and psychiatry exceed the limits of their usefulness in people's lives (and begin to do more harm than good), many people are coming to the realization that philosophical expertise encompasses logic, ethics, values, meaning, rationality, decision-making in situations of conflict or risk, and all the vast complexities that characterize human life.

People facing these situations need to talk in terms deep and broad enough to address their concerns. By getting a handle on their personal philosophies of life, sometimes with the help of the great thinkers of the past, they can build a framework for managing whatever they face and go into the next situation more solidly grounded and spiritually or philosophically whole. They need dialogue, not diagnosis.

You can apply this process in your own life. You can work on your own, though sometimes it helps to have a partner to converse with who can make sure you're not overlooking something or settling for rationalization over rationality. With the guidance and examples in this book, you'll be ready to discover the benefits of an examined life, including peace of mind, stability, and integrity. You don't need any experience in philosophy, and you don't have to read Plato's *Republic* or any other philosophy text (unless you want to). All you need is a philosophical turn of mind, which, since you've picked up this book and read this far, I'd say you have.

A PHILOSOPHY OF ONE'S OWN

Everyone has a philosophy of life, but few of us have the privilege or leisure to sit around and puzzle out the fine points. We tend to make it up as we go along. Experience is a great teacher, but we also need to reason about our experiences. We need to think critically, looking for patterns and putting everything together into the big picture to make our way through life. Understanding our own philosophy can help prevent, resolve, or manage many problems. Our philosophies can also underlie the problems we experience, so we must evaluate the ideas we hold to craft an outlook that works for us, not against us. You can change what you believe in order to work out a problem, and this book will show you how.

Despite its current reputation, philosophy doesn't have to be intimidating, boring, or incomprehensible. Much of what's been written on the subject over the years certainly falls into one or more of those categories, but at its heart, philosophy examines the questions we all ask: What is a good life? What is good? What is life about? Why am I here? Why should I do the right thing? What is the right thing? These are not easy questions, and there are no easy answers or we wouldn't still be mulling them over. No two people will automatically arrive at the same answers. But we all have a set of operating principles we work from, whether or not we are conscious of them and can enumerate them.

The great thing about having thousands of years of thinking to draw on is that many of history's wisest minds have weighed in on these subjects and have left insights and guidelines for us to use. But philosophy is also personal—you are your own philosopher too. Take what you can learn from other sources, but to arrive at a way of approaching the world that works for you, you'll have to do the thoughtful work yourself. The good news is that with the proper encouragement, you can think effectively for yourself.

And where do you find such encouragement? Here in this book, for starters. *Plato, Not Prozac!* offers you some of the fruits of philosophical practice. My fellow practitioners and I are not philosophers in the academic sense alone. Although many of us have Ph.D.s, teach in universities, and publish specialized articles, we do more than that: we also offer client counseling, group facilitation, and organizational consulting. We take philosophy out of purely theoretical or hypothetical contexts and apply it to everyday personal, social, and professional problems.

If you came to see me, I might discuss Kierkegaard's thoughts on coping with death, Ayn Rand's ideas on the virtue of selfishness, or Aristotle's advice to pursue reason and moderation in all things. We might look into decision theory, the *I Ching* (*Book of Changes*), or Kant's theory of obligation. Depending on your problem, we'd examine the ideas of philosophers most applicable to your situation, those you would be most inclined to relate to. Some people like the authoritative approach of Hobbes, for example, while others respond to a more intuitive approach, like Lao Tzu's. We might explore their philosophies in depth. But more likely you'd have your own philosophical outlook and be looking to express it more clearly. I'd act as a guide to elicit and illuminate your own ideas, and possibly to suggest some new ones to you.

What you'd find, having worked through your issue philosophically, is an open-minded, deep-seated, lasting way of facing whatever comes your way, now and in the future. You'd find this true peace of mind through contemplation, not medication. Plato, not Prozac. It requires clear, sharp thinking, but it isn't above your head.

Life is stressful and complicated, but you don't have to be distressed and confused. This book addresses human problems of everyday living. We are especially vulnerable when we are low on faith or confidence, as so many of us are who feel we can't find all the answers in religion or in science. Throughout this century, a widening abyss has opened beneath us as religion has retreated, science has advanced, and meaning has expired. Most of us don't see the abyss until we have fallen into it. Existential philosophers conduct extensive tours of it but cannot extricate many people from it. We need to glean the practical applications from all schools of philosophy to map our way through.

Philosophy is regaining its lost legitimacy as a helpful way to examine the world around us as our universe supplies us with new mysteries faster than either theology or science can reconcile existing riddles. Bertrand Russell characterized philosophy as "something intermediate between theology and science . . . a No Man's Land, exposed to attack from both sides." But the upside of that apt assessment of the downside is that philosophy can draw on strengths from both sides without having to absorb the dogmas or weaknesses in either.

This book draws on the greatest philosophers and philosophies throughout history and around the world to show you how to address the important issues in your life. It deals with the problems everyone faces, including handling love relationships, living ethically, coping with death, dealing with career change, and finding meaning or purpose. Of course, not every problem has a solution, but even if you can't find a solution, you need to manage the problem somehow so that you can get on with your life. Either way—solving or coping—this book can guide you. But rather than offering superficial New Age or pathology-oriented pseudomedical approaches, this book offers time-tested wisdom specifically geared to helping you live with fulfillment and integrity in an ever more challenging world.

PRACTICAL PHILOSOPHY

Philosophical counseling is a relatively new but rapidly growing field of philosophy. The philosophical practice movement originated in Europe in the 1980s, beginning with Gerd Achenbach in Germany, and started growing in North America in the 1990s. While *philosophy* and *practical* are two words not likely to be linked in most people's minds, philosophy has always provided tools for people to use in their everyday lives. When Socrates spent his days debating major issues in the marketplace, and when Lao Tzu recorded his advice on how to follow the path to success while avoiding harm, they meant these ideas to be used. Philosophy was originally a way of life, not an academic discipline—a subject to be not only studied but applied. It was only in the last century or so that philosophy became completely consigned to an esoteric wing of the ivory tower, full of theoretical insight but empty of practical application.

Analytical philosophy is the technical term for what probably pops into your mind when you think of philosophy. This is the field as defined academically. This kind of philosophy is mostly abstract and self-referential, with little or nothing to say about the world. It is rarely applicable to life. This approach is fine for universities. Basic study in philosophy should be part of every general education; a university without a philosophy department is like a body without a head. But in most academic fields of study where there's a pure branch of inquiry, there is also an applied branch. You can study pure or applied mathematics, theoretical or experimental sciences. While it is essential for any field of study to expand its theoretical frontiers, academic philosophy has lately overemphasized the theory, to the detriment of the practice. I'm here to remind you that the living wisdom of philosophy, which is concerned with real life and how to live it, predates the institutionalization of philosophy as mental gymnastics having nothing to do with life.

Philosophy is coming back into the light of day, where ordinary people can understand and apply it. Timeless insights into the human condition are accessible to you. We philosophical practitioners take them off the musty library shelf, dust them off, and put them into your hands. You can learn to use them. No experience necessary. You may want a walking tour of the territory before you set off on your own, and this book will give you that, along with a key to all the road signs you need to have a safe and fascinating trip on your own or with a friend.

By no means does benefiting from the wisdom of the ages hinge on having a Ph.D. or any such thing. After all, you don't need to study biophysics to take a walk, study engineering to pitch a tent, or study economics to find a job. Similarly, you don't need to study philosophy to lead a better life—but you may need to practice it. The truth about philosophy (and a well-guarded secret it is) is that most people can do it. Philosophical inquiry doesn't even require a degreed or certified philosopher, just a willingness to approach the subject in philosophical terms. You don't have to go out and pay someone—though you might enjoy and learn from the process with a professional—because with a willing partner, or even on your own, you can do it in your own home, or at a coffee shop or shopping mall for that matter.

As a philosophical practitioner, I'm an advocate for my client's interests. My job is to help my clients understand what kind of problem they face and, through dialogue, to disentangle and classify its components and implications. I help them find the best solutions: a philosophical approach compatible with their own belief system yet consonant with time-honored principles of wisdom that help in leading a more virtuous and effective life. I work with clients to identify their beliefs (and consider replacing unhelpful ones) and explore universal questions of value, meaning, and ethics. Working with this book, you can learn to do the same for yourself, though it may help to have another mind at work on the same subject.

Most clients come to me for reassurance that their actions are in accord with their own worldviews and rely on me to call attention to inconsistencies. The focus in philosophical counseling is on the now—and looking ahead to the future—rather than on the past, as in so much of traditional psychotherapy. Another distinction is that philosophical counseling tends to be short-term. I see some clients for just one session. Typically I'll work with someone several times over a period of a few months. The longest I've counseled one person is a year or so.

A philosophical counseling session involves more than simply matching problems with sound bites from philosophical literature, though a simple aphorism can sometimes cut through a knotty problem. It is the dialogue, the exchange of ideas itself, that is therapeutic. This book will give you all the information you need to clarify your own philosophy, as well as guidance on how to conduct your inner deliberation or dialogue with a friend. I'll show you how to be radical enough to contemplate every option, but prudent enough to choose the right one.

PLATO OR PROZAC?

Before relying solely on philosophy to manage a problem, you must make sure it is appropriate to your situation. If you are upset because you have a stone in your shoe, you don't need counseling—you need to remove the stone from your shoe. Talking about the stone in your shoe will never stop your foot from hurting, no matter how empathic a listener you find and no matter what school of therapy he or she subscribes to. I refer people with problems I suspect to be physical in nature to medical or psychiatric professionals. Some people may not benefit from Plato, just as others may not benefit from Prozac. Some may need Prozac first, then Plato later, or Prozac and Plato together.

Many people who seek out philosophical counselors have already been in therapy but found it ultimately unsatisfying, at least in some regard. People can be harmed by psychological or psychiatric treatment if their root problem is philosophical and the therapist or doctor they seek out doesn't understand that. A feeling of desperation can arise if you begin to feel that no one will be able to help you with your problem because you are not being listened to or heard clearly. Inappropriate therapy wastes time (at best), and can make your problem worse.

Many people who ultimately rely on philosophical counseling have first benefited from psychological counseling, even if they find it isn't enough on its own. Your past certainly conditions you and informs your habitual way of looking at things, so examining your past can help. Understanding your own psychology can be a useful preparation for cultivating your own philosophy. We all carry psychological baggage, but getting rid of the excess may require philosophical counseling. The idea is to travel as light as possible. Knowing yourself—a goal approached differently in psychological and philosophical counseling—is not about memorizing the Encyclopedia of You. Dwelling on each exquisite detail adds to your baggage rather than lightening your load.

Many people who wouldn't touch psychotherapy with a ten-foot pole find the idea of talking to someone about ideas and worldviews both appealing and acceptable. Whether or not psychological therapy is a stepping stone for you, if you are curious, speculative, pensive, analytical, and articulate you can benefit greatly from philosophical counseling. In fact, anyone with an inquiring mind is ready for the examined life that philosophers take as their unifying purpose.

THERAPY FOR THE SANE

Philosophical counseling is, in the words of my Canadian colleague Peter March, "therapy for the sane." To my mind, that includes just about all of us. Unfortunately, too much of psychology and psychiatry have been aimed at "disease-ifying" (that is, medicalizing) everyone and everything in sight, looking to diagnose each person who walks in the door and find what syndrome or disorder could be the cause of their problems. On the flip side, a lot of New Age thought takes as a premise that the world (and everyone in it) is just as it should be or is meant to be. While we should generally expect to be accepted despite the variety-pack of idiosyncrasies and flaws we all come with, and while there is no reason to see those flaws as somehow abnormal (perfection is what is abnormal), there is also no reason to see change as beyond our reach. When Socrates declared that the unexamined life is not worth living, he was arguing for constant personal evaluation and striving for self-improvement as the highest calling.

It is normal to have problems, and emotional distress is not necessarily a disease. People looking for a way to monitor and manage a world that is growing ever more complex don't need to be labeled with a disorder when they are actually treading a time-honored path to a more fulfilling life. In this book, you'll see specifically how to apply philosophy when you are facing moral dilemmas; professional ethical conflicts; difficulty reconciling your experience with your beliefs; conflicts between reason and emotion; crises of meaning, purpose, or value; quests for personal identity; a search for parenting strategies; anxiety over career change; inability to achieve your goals; midlife changes; problems with relationships; the death of a loved one or your own mortality. I've selected for detailed examination the most common life situations that bring people face-to-face with their own philosophies. No matter what your particular concern, you can apply the underlying techniques and insights.

As it guides you in using philosophy on your own, this book does much more than suggest that you "take two aphorisms and call me in the morning." It is a practical guide to life's most common struggles. It provides a quick survey of philosophy for those who never took (or can't remember) Philosophy 101. But it is also a serious consideration of ways we can all live with more integrity and satisfaction. It takes on the major questions everyone faces in life—and offers the answers given by some of the great minds of all time, as well as strategies to help you arrive at the answer that will have the most meaning for you: your own.

OVERVIEW

So that you know what you're in for, the following is a brief description of where to find what in this book.

Here in Part I, I introduce philosophical practice, ways of using philosophy to help yourself, and the limits of self-help. After this chapter about what's gone wrong with philosophy as it is used (or not used) in the real world—and what's going right with it lately—chapter 2 looks at strengths and weaknesses of psychology and psychiatry and compares various kinds of therapies. Chapter 3 presents the five steps of the PEACE process for managing problems philosophically. Chapter 4 provides a brief overview of some philosophers whose ideas are relevant to my counseling practice, to give you some historical perspective.

Each of the chapters in Part II focuses on one of the more common problems brought to philosophical counselors, shows how philosophy addresses them, and guides you in applying philosophical thinking to your own particular situations. Case studies are interspersed with explanations of the major schools of thought and specific important philosophical theories to provide a range of options for each common situation. These are the tools you need to examine your own life.

Part III broadens the perspective beyond philosophical counseling and looks at how philosophy is practiced with groups and organizations. The term *philosophical practice* refers to three types of professional activities: counseling individual clients, facilitating various kinds of groups, and consulting to various kinds of organizations. Hence a philosophical counselor is one type of philosophical practitioner. Some practitioners specialize in one type of practice; others branch into more than one type. While this book focuses on counseling, the other kinds of practice are also important and worthy of mention here. Individuals working in groups can derive personal gain from such encounters, and organizations can benefit from the same kind of self-examination and philosophical clarification as do individuals and groups.

Finally, in Part IV, the list of additional resources gives you lots of further information. Appendix A lists a hit parade of useful philosophers and their greatest works. Appendix B lists organizations for philosophical practice in America and abroad. Appendix C is a national and international directory of philosophical practitioners. Appendix D offers a selection of further reading on philosophical practice and related topics. Appendix E shows you how to use the *I Ching*, a perennial source of philosophical wisdom that you can consult on your own.

This book is much more didactic than a typical one-on-one philosophical counseling session would be. With a philosophical practitioner, a session might run one of three ways. You might discuss your problem in generic terms, without mention of any particular philosopher or philosophy. This is the kind of conversation you're more likely to have with your friends, partner, family, bartender, or cab driver— and sometimes it is the best approach. You're thinking for yourself, using critical and analytical skills, calling upon your insight into yourself, and conversing philosophically without self-consciously trying to be philosophical.

Another common path in a counseling session is for the client to ask specifically for some philosophical instruction. In this variation, you may have reinvented a philosophical wheel and be reassured to know that someone else has mapped out that territory already. In the likely event that you haven't covered all the bases yourself, you could learn from those who have gone before you. Understanding a bit of the formalized school of thought may help you connect your own dots or fill in some blank spaces, but your counselor usually won't give you a whole disquisition on each topic—unless you ask for it.

A third, more hard-core alternative is for those who have worked through problems this way but still have further interest. This gets more involved and may mean you get referred to another practitioner or explore bibliotherapy—actually tackling some philosophical texts. You may have been helped by a Buddhist insight and want to learn more about practicing Zen. Or you may have been helped by an idea of Aristotle's and want to delve more deeply into his system of ethics. This kind of work can help gird you for future issues even more thoroughly than the experience of working through a specific problem, but it is just one option and certainly won't be right for everyone.

While my counseling practice involves all three approaches, this book primarily illustrates the second one. You are not expected to know specific philosophies—that's my job. Your job is to bring the problem and your willingness to inquire philosophically. The dialogue—internal or external—that results will help you interpret, resolve, or manage whatever your issues are. I'm not a medium who will put you in contact with dead philosophers, but I am your guide to their ideas, systems, and insights. Once you've been introduced to them, they'll serve you well in handling what comes your way.

We've found that neither science nor religion answers all our questions. The philosophical psychotherapist Victor Frankl saw this as leading to an "existential vacuum" from which ordinary people needed a new path out:

> More and more patients are crowding our clinics and consulting rooms complaining of an inner emptiness, a sense of the total and ultimate meaninglessness of their lives. We may define the existential vacuum as the frustration of what we may consider to be the most basic motivation force in man, and what we may call . . . the will to meaning.

Frankl used the phrase "will to meaning" to parallel two of the central ideas of psychology: Adler's "will to power" and Freud's "will to pleasure." But as Frankl foresaw, there was something still deeper at the heart of most people's central problem, and existing medical, psychological, and spiritual treatments were not going to be enough to relieve it. We once directed our questions about meaning and morality to some traditional authority, but those authorities have crumbled. A great many people are no longer satisfied with passively accepting the dogmatic dictates of an inscrutable deity or the impersonal statistics of an imprecise social science. Our deepest questions remain unanswered. Worse, our beliefs go unexamined.

The alternative is the practice of philosophy. It is time for a novel way of looking at things, and the novel way described herein is actually an ancient way, long forgotten but lately recollected. As we enter the new millennium, we have come full circle.

"We shall not cease from exploration
And the end of all our exploring
Will be to arrive where we started
And know the place for the first time."
—T. S. ELIOT

2

Therapy, Therapy Everywhere, and Not a Thought to Think

*"Since some strange views have been put forward,
by error or with some ulterior motives . . . philosophers have
been compelled to assert the truth of the manifest or to deny
the existence of things wrongly imagined."*
—MOSES MAIMONIDES

*". . . the notion of mental illness is used today chiefly to obscure
and 'explain away' problems in personal and social relationships,
just as the notion of witchcraft was used for the same purpose
from the early Middle Ages until well past the Renaissance."*
—THOMAS SZASZ

America has become a therapeutic society. Or, rather, therapized. More practitioners hang out their shingles every day, and what it means about your qualifications (if any) to call yourself a therapist varies according to state law—not substance. Throw out "My therapist says . . ." at a cocktail party, and you won't be able to get a word in edgewise for the rest of the conversation as everyone else jumps in with what their therapists say. When President Clinton convened his Cabinet to address the scandal shadowing his office, participants described it to the *Washington Post* not as a strategy meeting or a political rally, but as an "encounter session." Bartenders, taxi drivers, your best friend, your mother, and passing acquaintances are always on hand with some nugget of quasi-psychological advice to remedy your every travail. "Self-help" bookshelves seem to go on for miles at your local superstore. Television talk shows pioneered bringing instant insight of the most superficial sort to every facet of human behavior. Even now, when the script always seems to call for inciting a fistfight among panel participants, no doubt a therapist of some kind will be brought in just before the credits roll to pay lip service to working it out in a more civilized fashion. It's a wonder anyone survives the month of August, when all the therapists are on vacation.

This has been going on for decades now, but we don't seem to have learned much, since the demand for assistance with our mental and emotional health continues unabated. High-quality psychological counseling or psychiatric care can provide valuable assistance and workable solutions to many kinds of personal turmoil. But both areas (like all areas) are limited in scope and so cannot provide lasting or complete results for everyone, not even for many of the people who initially derive important benefits. It isn't enough.

Philosophical counseling also cannot address everything, and I need to be able to refer my clients for psychological or psychiatric care sometimes, in addition to, instead of, or before working philosophically. But philosophical counseling goes a lot further in providing workable long-term approaches to the most common problems people seek counseling for, and can fill in some of the gaps left by other kinds of counseling. This chapter explores the usefulness and limits of psychology and psychiatry and shows how philosophical counseling fits in.

A GAME OF CHESS

The metaphor of a chess game (inspired by my philosophical counseling colleague Ran Lahav) illustrates the differences between psychological, psychoanalytic, psychiatric, and philosophical approaches to counseling. Imagine you're in the middle of a chess game, and you've just made a move.

Your psychotherapist asks you, "What made you make that move?" "Well, I wanted to capture the castle," you say, unsure what she's getting at. But she'll keep asking questions to find the supposed psychological cause of that move, sure that there's some explanation beyond "I wanted to capture the castle," and you may end up telling your whole life story to satisfy her assumptions. A once-popular but now strongly criticized psychological theory would have suggested that your present aggressive behavior—wanting to capture the castle—stems from some past frustration.

Your psychoanalyst would ask you the same question: "What made you make that move?" When you answer, "Well, I wanted to capture the castle," he'll follow up with, "Very interesting. Now what made you say that that was what made you make that move?" You might again have your entire life story dragged out of you, or at least the chapters on your childhood. If he's still not satisfied, he might posit some reasons you had but were unconscious of, going back to early infancy. A still-current but now strongly criticized psychoanalytic theory would have suggested that your possessive behavior—wanting to capture the castle—stems from your repressed insecurity at having been weaned from the breast.

Your psychiatrist would also ask you, "What made you make that move?" Again you reply, "Well, I wanted to capture the castle." Now your psychiatrist consults her latest edition of the *Diagnostic and Statistical Manual* (DSM) until she finds the personality disorder that best describes your symptoms. Ah, there it is: "Aggressive-Possessive Personality Disorder." A still-current but increasingly criticized psychiatric theory would have diagnosed your behavior as the symptom of a brain disease, and you would be drugged to suppress this so-called symptom.

By contrast, your philosophical counselor might ask you, "What meaning, purpose, or value does this move have for you now?" and "What bearing does it have on your next move?" and "How would you assess your overall position in this game, and how might you improve it?" The philosopher looks at the move not merely as an effect of some past cause, but as something significant in the present context of the game itself, as well as the cause of future effects. He'll recognize that you have a choice about the moves you make and will see the cause of the move as perhaps relevant but certainly not the whole story.

I think it is much more wholesome to be living your life rather than constantly digging up its roots. If you did that continually to even the hardiest plant, it would never thrive, no matter how much TLC you otherwise showered upon it. Life is not a sickness. You can't change the past. Philosophical counseling starts from there and goes forward to help people develop a productive way of looking at the world, and so a comprehensive plan for how to act in it day-to-day.

THE PHILOSOPHY/PSYCHIATRY/PSYCHOLOGY SPLIT

Philosophy and science were once one and the same occupation. Aristotle studied astronomy and zoology as well as logic and ethics. Robert Boyle (as in Boyle's law: the volume of a gas, at a constant temperature, is inversely proportional to the pressure) would have put "experimental philosopher" at the top of his résumé. Laws of motion were discovered by natural philosopher Sir Isaac Newton, biological evolution by natural philosopher Charles Darwin. Philosophers like them were engaged in testing and measuring the world around them, a process that began as an extension of the kind of "How does the world work?" questions most philosophers asked. Before the scientific revolution in the seventeenth century, more allied these approaches than separated them.

Ultimately science and philosophy followed divergent roads, and Western medicine—after centuries in the hands of charlatans, barbers, phrenologists, and snake-oil vendors—allied itself with science. Psychiatry developed as a branch of primitive medicine in the eighteenth century and really established itself during the twentieth, in the wake of Freud. Medicine is still a balance of science and art: CAT scans and bedside manner, chemotherapy and visualization techniques, electrocardiograms and second opinions. Freudian psychoanalysis and all its breakaway forms developed by his dissenting disciples (Jung, Adler, Reich, Burrow, Horney, and others) has become more like a schismatic religion than anything else. Freudian and Jungian psychoanalysts are as divided and mutually hostile as Ultra-Orthodox and Reconstructionist Jews, Catholic and Protestant Christians, or Sunni and Shi'ite Moslems. You don't need to be a medical doctor to be a psychoanalyst; you do need to cleave—at all costs—to a particular doctrine.

Then again, Freud's philosophy of psychiatry was that all mental problems (what he called neuroses and psychoses) would eventually be explained in terms of physical ones. In other words, he thought that every mental illness is caused by a brain disease. And this is exactly where modern psychiatry has gone. Any conceivable behavior can end up in the DSM, where it is diagnosed as a symptom of a supposed mental illness. Although most of the so-called mental illnesses in the DSM have never been shown to be caused by any brain disease, the pharmaceutical industry and the psychiatrists who prescribe their drugs are committed to identifying as many "mental illnesses" as they possibly can. Why? For the usual reasons: power and profit.

Consider this: In 1952, the DSM-I listed 112 disorders. In 1968, the DSM-II listed 163 disorders. In 1980, the DSM-III listed 224 disorders. The latest edition, the 1994 DSM-IV, lists 374 disorders. In the 1980s, psychiatrists estimated that one in ten Americans was mentally ill. In the 1990s, it was one in two. Soon it will be everyone—except, of course, for the psychiatrists. They find "mental illness" everywhere—except in the laboratory—and prescribe as many drugs as your insurance company will pay for.

While there are certainly some people who need to be medicated or psychiatrically incarcerated to prevent them from harming themselves and others, that number is nothing like one American in two, or ten, or a hundred. For the most part, personal unhappiness, group conflict, gross incivility, shameless promiscuity, epidemic crime, and orgiastic violence are products not of a society that is mentally ill, but of a system that—through lack of visionary statesmanship and philosophical virtue—has allowed and encouraged society to become morally disordered. Though philosophers have remained mostly silent on this issue, philosophical practitioners can help restore moral order—and with it "mental health"—to our thoroughly demoralized citizenry. Moral order isn't a drug, but it does have wonderful side effects.

Psychology didn't emerge as a field of study in its own right until 1879, when Wilhelm Wundt opened the first psychology laboratory. Before that, the kinds of observation and insight we associate with psychology were the province of philosophers. Even after psychology came into its own, philosophy and psychology remained twin disciplines into the twentieth century. William James, hailed as a great thinker in both fields, held a joint chair in philosophy and psychology at Harvard into the early 1900s, while Cyril Joad had a similar appointment at the University of London into the 1940s. But the fields have diverged over the last century, with psychology moving out of the humanities wing of the academy and settling in with the social sciences instead. Despite his foothold in the philosophical realm, James was a leading advocate for making psychology a science: "I wished, by treating Psychology like a natural science, to help her to become one," he wrote.

With the advent of behavioral psychology, the split was complete. Behavioral psychologists like John Watson and B. F. Skinner took their questions about human nature into the laboratory and ran experiments on them. That's a far cry from the approach epitomized by Rodin's *The Thinker* (chin in hand, elbow on knee, lost in thought) that philosophers stereotypically favor. But whether you develop your ideas by having verbal duels with Socrates or running rats through mazes, the questions you are ultimately asking are basically the same: What makes a human being tick? Is it rational will or conditioned response? If it's both, how do they interact?

Philosophers have always been observers of human nature, which sounds like the job description of a psychologist. Any philosophy of humanity would be incomplete without psychological insight. Psychology, too, fails when it is devoid of philosophical insight, and both fields are the poorer for having become so bifurcated. Some areas of philosophy, like logic, stand far apart from psychology. But in the main, philosophy relies on observation, sense data, perception, impressions—all of which cross into psychological territory. When we look at the world, we don't necessarily see clearly what is in front of us—physiological quirks and subjective interpretations almost always play a part. That interpolation—the difference between object and experience—is psychology, and no philosophical outlook holds up without it.

Behavioral psychology and its central stimulus-response theory regard a person as a kind of machine that can be conditioned or programmed for any desired result—you just need to find and use the right stimulus. (S-R theory is what Pavlov was confirming when he got dogs to salivate at the sound of a bell, having trained them by ringing a bell just before putting a plate of food in front of them.) But much is elided by that hyphen between S and R. All the rich, important parts of psychology—and of humanity—are disregarded in distilling all actions to simple cause and effect. Thinking of a human as no more than a creature responding in controllable ways to specific stimuli diminishes our humanness. It disregards the psyche—the ostensible subject of study in psychology. We are much more than just our conditioning; there is more to our lives than a series of set responses. The problem is that much of modern psychology—psychology as a science—is descended from or influenced by behavioral psychology and its attendant impoverishment of human experience.

Applying the scientific method yields some important information about humans and how they work. But though it may pick out threads of insight, psychology will never reveal the whole complex tapestry of human nature. For example, proper scientific method notwithstanding, behavioral psychology will never yield a system of ethics, one of the key components of human life—and a subject an entire vein of philosophy is devoted to. If eliciting an action is just a matter of finding the right stimulus, humans are reduced to doing everything they do simply to get a reward or avoid a punishment. (The stimulus is either carrot or stick.) Under those conditions, is there any such thing as doing a good deed? Would it be possible to do what is right simply because it is the right thing to do, and not do the wrong thing simply because it wouldn't be right?

Behaviorists would say that if they zapped you with an unpleasant electric shock each time you helped a little old lady across the street, you'd soon cut out the good Samaritan routine. They'd also claim they could get you to push her as you whizzed past on your own way across the street if they gave you the right reward each time you did so. In this way, behaviorists make humans too shallow, ignoring our rich inner mental lives. We are much more varied and complex than the rats who will manically push the lever that usually delivers food long after the treats have stopped coming. (We all have some times when we don't behave much more intelligently than those rats, but that's another story.)

One of the gifts humans have is the ability to provide our own internal stimuli. Sometimes we promise ourselves a dish of ice cream after an unpleasant chore, and then we are using what we've learned from the behaviorists. But we can also motivate ourselves with a sense of honor or duty or service, topics the philosophers have dwelt upon but that are mainly beyond the scope of experimental psychology. That's why Arthur Koestler dubbed it "ratomorphic psychology": the experimenters ended up learning a lot about rodents, but the take-home lessons for humans are limited, and certainly can't touch the big questions of our existence.

All scientists work with sets of "observables"—the things they study. For example, astronomers have galaxies, stars, and planets; chemists have atoms, molecules, and so on. The scientists' job is to make and record observations about whatever phenomena they are observing, to posit theories to explain why whatever it is behaves as it does, and then to test those theories by conducting experiments. In the social sciences, the set of observables is not quite physical or directly measurable. That results in major philosophical differences between physical (or natural) and social sciences, and means you'll never find one all-encompassing science division at a university. In the social sciences, researchers impose their own worldviews on whatever they're observing—which is why even brilliant minds like Margaret Mead's came under fire for drawing faulty conclusions based on subjective (perhaps even unconscious) bias. Physical scientists may run into similar problems, but the effect on the outcome of research will be mitigated by the more concrete, objective nature of the items under observation.

In psychology, the set of observables consists of the psyche. How do you observe that? What is it, even? Neuropsychology observes the brain, which is measurable, at least to some degree. But generic psychology observes the mind. Since the mind, or the psyche, does not have physical characteristics, all observations are indirect and all conclusions are more subjective and less certain than they are in the physical sciences. Even in the physical sciences, where we have the benefit of direct observation, our information is imperfect. As many unanswered questions as we have about the mind, we have nearly as many about the touchable, weighable, dissectable brain. So imagine how much easier it is to go wide of the mark in the social sciences, like psychology, when the whole business is much more abstract and we have nothing concrete to observe.

Nonetheless, the relatively recent scientization of psychology combined with the perennial human need for dialogue has resulted, in the twentieth century, in the unprecedented growth of the psychological counseling industry. When the first psychologists began to counsel human beings, academic psychologists accused them of heresy, apostasy, and the like. "Psychological counseling isn't psychology," said conventional wisdom. But within a few decades, counseling psychologists vastly outnumbered all other kinds of psychologists put together.

Counseling psychologists have a virtual monopoly on government licensing of talk therapy, which is why your health insurance will help pay for the cost of seeing them, even though they aren't medical doctors. If your primary care physician can refer you to a psychological counselor, who is not a doctor but whose services are paid for by health insurance, then you ought to be able to get a referral to a philosophical counselor too.

Social scientists must rely on statistics in making their measurements. While statistics may tell you a lot about a group, they tell you nothing about an individual, and here psychologists run into another wall. We are generally right to accept reliable statistics but also to insist that they do not describe us as individuals. Psychological data show that, on average, girls have stronger verbal skills than boys, and boys have stronger spatiotemporal skills than girls. But that doesn't mean much to the boy who just scored 800 on the verbal portion of the SAT or to the girl who's pitching for a top Little League team. What have we really learned that we can use? (That's the pragmatic voice of William James, testing an idea for "truth" by seeing if we can use the information.) Do boys need more reading classes? Do girls need more sports? Can we effectively teach boys and girls together in the same groups? We can examine the statistics and try to decide what will work for large groups. But if you're a parent looking for the best situation for your child, you'd do just as well to follow your experience and intuition.

You may think philosophy couldn't design the best lesson plan either. But with "facts" of limited usefulness, it is educational philosophy that ultimately guides those decisions. Most boys and girls are in coeducational schools because we have a philosophical commitment to equality of opportunity (even if that means small and perhaps temporary differences in reading scores). In addition, educators choose the method of reading instruction that they think is most effective—and the differences there are philosophical too. Too many American schools are graduating illiterate students, partly because of a systemic philosophical commitment to teaching methods that virtually guarantee illiteracy (e.g., whole word) instead of those that guarantee literacy (e.g., phonics). Since we know how to teach children to read, as well as how to teach them not to read, the method of education we choose depends upon nothing but our philosophy of education.

No matter what kind of scientist you are, the questions you are trying to answer are "How does this thing work?" and "Why does it do what it does?" Scientists look for cause and effect (a point the behaviorists simplified to an extreme). Psychologists, then, are asking, "How do people work? Why do they do what they do?" And a psychological therapist is asking, "How does this person work? Why does she do what she does?" The therapist is seeing effects (e.g., anxiety, depression) and looking for causes (e.g., a negative relationship with a partner, alcoholic parents). That may be the scientific way to look at things, but when applied to an individual, psychological therapy runs into two big problems.

Cause and Effect

The first is a fallacy that philosophers call *post hoc ergo propter hoc.* In case you're not up on your Latin, it means that because one event happened before another, the earlier event caused the later one. That may be true in some cases: you stub your toe, then shout "Ouch!" But it is not necessarily or always true. If your parents hit you as a child, and you now have trouble controlling your anger, you can't conclude that the one caused the other. Perhaps it did. But it may well be irrelevant. Even if it were causal, there may be many other contributing factors. To go back to stubbing your toe, it would be obvious that not wearing shoes, and leaving junk on the floor, and rushing across a room to answer the phone in the dark are all causal events. But in the complicated stew of your past, can every ingredient really have a bearing on the issue at hand? Did your mother's rages teach you to be explosive when you're angry? Or was it your father's coldness? Both? Neither?

Casting back over the events of your life looking for causes of your current difficulty is further problematic because there may be connections that you can't see. And your memory isn't perfect, so some important facts will be forgotten and many irrelevant details remembered. What if this mining of the past turns up only fool's gold—but you and your therapist take on as if you've just tapped a major vein of ore? At best you're going to waste a lot of time on irrelevancies and wrong turns, even if you eventually arrive at your destination.

With no neat laws to guide you, as a chemist or physicist has, how do you know what makes what happen? If in principle anything could have caused anything else (as long as one preceded the other), the danger is that once you have a theory, you just pick out the items that fit your theory and disregard the rest. As Abraham Maslow finely pointed out, if the only tool in your tool box is a hammer, a lot of things start looking like nails.

Even if psychology became a precise instrument, what good would finding the causes of your current discomfort do you? Would having a label, or being classified as "uncomfortable," make you feel any better? Knowing you have a cavity doesn't stop your pain—getting it filled does. Understanding that you have a headache because you were beaned by a fly ball doesn't make you feel any better, but an aspirin might. Yes, you should improve your dental hygiene and polish your outfielding skills to avoid similar pains in the future. Yes, some people will be relieved by discovering the source of their psychic pain through psychology, and others will be able to see a course of action that will bring relief once they understand the cause. But for many more, pinpointing the cause won't be enough to help them. They'll spend months or years digging until they hit pay dirt—and then their response is likely to be, "So what?" Knowing the cause of their psychic pain but having no avenue to reduce it will make some people feel even worse. Knowing you're depressed because your marriage is falling apart might only increase your depression since you can't go back and change the past.

The Medical Model

The second major problem with psychiatric and psychological therapies is that they mimic the medical model. They are licensed by states as though they were medicine, and health insurance covers them (at least partially). Medical doctors are trained to diagnose and treat physical illnesses. Psychiatrists and psychologists are trained to treat "mental illnesses." That term is in quotes because it is a metaphor. But the metaphor is increasingly mistaken for a reality. An old psychiatric joke illustrates the point: patients who came early for their appointments were diagnosed as anxious; patients who came late, as hostile; patients who came on time, as compulsive. This joke was told by psychiatrists themselves, who knew full well the difference between literal and metaphoric illness. But it's not funny any more, because that difference has been blurred by what Thomas Szasz calls "the myth of mental illness."

Medical problems are often called "syndromes." Hosts of syndromes have been observed, documented, researched, and understood. For example, Down's syndrome is caused by a specific genetic sequence, while Tourette's syndrome is a specific brain dysfunction that manifests as agitated but not dangerous gestures and vocalizations. Acquired immune deficiency syndrome (AIDS) is caused by a retrovirus (HIV) that attacks and disables the immune system. But what about diagnoses like "Gulf War syndrome"? What does that mean? Apparently it means that some people who served in the Gulf War aren't feeling well. No one knows (or no one is saying) whether they were exposed to biological agents or chemical toxins, whether their problems are medical or psychological or both. Making a diagnosis like Gulf War syndrome sounds very scientific, but it reveals no new or useful information about the problem. Similarly, consider "sudden infant death syndrome" (SIDS). Unfortunately, some infants do die unexpectedly in their cribs. It used to be called "crib death." Now it has a much fancier name, which still reveals precisely nothing about the problem. So simply calling something a syndrome doesn't guarantee that we know what we're talking about, even when there's something medically (that is, physically) wrong.

This is not news to philosophers. Here's a famous "explanation" about opium that dates from the Middle Ages. The question was, why did opium (when used medicinally) put people to sleep? The answer given by doctors of the day (some of whom, I'm sorry to say, were probably philosophers) was that opium put people to sleep because of its "dormitive properties." Everyone nodded wisely and accepted for years that this really explained something. But it explains nothing. *Dormitive* comes from the Latin verb *dormire*, "to sleep." To explain that opium puts people to sleep because of its dormitive properties is to explain that opium puts people to sleep because it puts them to sleep. Not very scientific after all—merely a circular explanation.

Now what happens when we apply circular definitions of literal physical illnesses to metaphoric mental ones? We get a zoo of so-called disorders. Do you have an unresolved emotional problem stemming from a past unpleasant experience? In the DSM, it becomes a mental illness: post-traumatic stress disorder. Is your child having a problem learning arithmetic? There's a good chance it's because her teacher doesn't know any or because current teaching methods claim that the right answer to 2 + 2 = ? is whatever number makes the student feel good, but in the DSM, it becomes a mental illness: developmental arithmetic disorder. Are you disappointed because you didn't win the latest lottery? In the DSM, it too becomes a mental illness: lottery stress disorder. Would you refuse psychiatric treatment for yourself or your child if you were confronted with this kind of diagnosis? In the DSM, your refusal itself becomes a mental illness: noncompliance with treatment disorder.

This would be great if it were science fiction or comedy. But it masquerades as serious science today. In 1987 the American Psychiatric Association voted in attention deficit hyperactivity disorder (ADHD) as a mental illness—science by ballot. In that year, half a million American children were diagnosed with ADHD. In 1996 it was estimated that 5.2 million children—10 percent of American schoolchildren—were diagnosed with ADHD. The "cure" for this "epidemic" is Ritalin, whose production and sales—and nightmarish side effects—have skyrocketed. This is very good for the drug business; not so good for the children. There is not one shred of medical evidence that ADHD is caused by any specific brain disease, but that is the claim that justifies pronouncing millions of American schoolchildren mentally ill, drugging them by coercion, and recording these "diagnoses" of "mental illness" on their permanent records.

Why are normal, healthy, inquisitive—and at times rambunctious—children having trouble paying attention in school? ADHD is one possibility; there are many others. It could also be because of no motivation, no discipline, no subject matter left to study, no standards that demand any learning, no tests that evaluate any knowledge, incompetent teachers, and indifferent parents. It could be because mandatory standards have been replaced with mindless slogans, and there is no moral authority at home or at school to inculcate virtues in these children. The educational system has been transformed from a path of learning to a minefield of debilitation—with psychology and psychiatry as willing accomplices. These accomplices have similarly infiltrated and colonized the justice system, the military, and the government. Is it any wonder that people are now turning (or returning) to philosophy?

It's easier for us to daydream. If your child doesn't pay attention in school, he's got attention deficit hyperactivity disorder. And if you complain about this kind of diagnosis, you've got attention deficit disorder denial disorder.

One problem with calling this approach scientific is that these so-called disorders are not tested according to any scientific criteria. The declaration or supposition that something exists without evidence to back it up is what philosophers call "reification." Psychiatrists and psychologists are experts at reifying syndromes and disorders: dreaming them up, then finding symptoms in people and calling it evidence that the illness exists. Whatever benefit is to be gained from grouping symptoms like that, there is also a huge drawback: it deadens the powers of inquiry, making us think we have answers when we don't. Feeling unhappy for no reason? Ah, that's unmediated depressive syndrome, see it all the time. Like to drum your fingers on your desk? You've got percussive digital disorder. Here's pragmatism again, to ask, "Where does any of this get us?"

So what if you are a woman who loves too much, or you have Peter Pan syndrome, or your husband is from Mars? Any self-help book worth its salt holds out the promise of self-improvement and of bettering your life in general. But psychology as it is currently constituted—and popular psychology especially—does not have the tools to help you apply the narrow insights you learn about yourself to the big picture of your life. Psychology can take you only so far, no matter what the cover of the latest best-seller promises. To integrate every conceivable insight (psychological insights being just one kind) into a coherent, workable outlook on and approach to life—a personal philoso-phy—you need . . . philosophy. Many of my clients have already been through extensive psychological work, and although many have derived preparatory benefits from it, none have found that it alone could bring a sense of inner peace to their lives. The people who do find psychology to be the secret of their success don't find their way to my office, however.

There's nothing inherently wrong with relying on theories if they are helpful to people (that's what philosophical counselors do, after all), but doing it in the name of science is misleading. Therapy, or counsel-ing, is primarily an art. Too much about it is subjective to place it in the objective realm of laboratory science. And anyway, must we be labeled with some kind of syndrome or disorder just because we face emotional, intellectual, psychological—or, yes, even philosophical—challenges? Of course not.

Neither can psychiatry adequately address the everyday problems most people need to talk about. The post-Freudian emphasis on bio-logically based illnesses with mental or emotional symptoms—and the prescription medications that can control them—makes psychiatry rel-evant to a tiny fraction of the populace. Those who are dysfunctional by reason of physical illness entirely beyond their control—such as manic-depressives—are helped by medication. For handling that kind of problem, make your first stop a psychiatrist's office. But if your problem is about identity or values or ethics, your worst bet is to let someone reify a mental illness and write a prescription. There is no pill that will make you find yourself, achieve your goals, or do the right thing.

If your root problem is philosophical, nothing on your pharmacist's shelves is going to give you lasting relief. Americans have a weakness for the quick fix. We have trusted in technology to improve our lives and provide an easy answer to everything. Our society also eagerly embraces excuses that reduce personal responsibility for anything undesirable. Even smokers who kept up their pack-a-day-plus habits after the haz-ards were common knowledge are now suing tobacco companies for giving them lung cancer, as if the companies alone were at fault. How better to free ourselves from tiresome burdens than to consider any kind of unhappiness or misbehavior a disease, something genetic or biological or environmental, and therefore beyond all our control? On top of that, drugs are cheaper and more plentiful here than anyplace else in the world.

All that conspires to align us with the psychiatric view of things. But hewing to this point of view gives you only a hollow sense of not being to blame and a false sense of hope about easy answers. Because there are no easy answers. The only way to have a true, lasting solution to a current personal problem is to work at it, resolve it, learn from it, and apply what you learn to the future. That's the focus of philosophical counseling, distinct among the countless types of therapy available.

FOUR FACES OF DEPRESSION

To understand how psychology, psychiatry, and philosophy look at the same thing differently, and the effect it has on treatment, let's look at four different ways of understanding depression. Each lens will give you clear vision in some cases but distort what you are viewing in others. If we always used the proper lens in the proper way at the proper time, we'd have the best of all worlds for helping someone deal with a problem as cleanly, efficiently, and lastingly as possible. But too often a counselor uses only one lens or fails to switch lenses or refer to someone with a different lens as the case warrants.

One possible cause of depression is that there is something wrong with the brain—a genetic problem causing neurochemical transmitters to be produced and released in a way that interferes with acceptable ranges of brain function. This type of depression is a physical illness, with all kinds of other consequences. Another kind of depression is caused by an induced brain state—that is, it is biological but not genetic. It could be the result of substance abuse—that is, a side effect of amphetamines or of a depressant like alcohol. This kind of depression indicates a physical or psychological dependency. A third typical cause of depression is unresolved childhood trauma or another problem in the past, which is a distinctly Freudian (and generally accepted) view and is a psychological but not a medical problem. A fourth kind of depression results from something acute happening in one's current life. The something might be a professional crisis, an impending personal or financial problem like divorce or bankruptcy, or a moral or ethical dilemma. Here there's nothing physical or psychological causing the depression; brain chemistry, substance abuse, and childhood trauma are not the culprits.

In the first two instances, a person needs medical attention. Psychiatry is very good in this kind of case, as medication may well control the symptoms. But drugs can't cure the underlying problem—though perhaps genetic engineering one day will—so talk therapy would still be indicated. In the third and fourth instances, talk therapy would be the apt prescription. For unresolved past problems, psychology has a lot to offer, though philosophical counseling would also be helpful, either instead of or after psychological counseling. But in the fourth scenario—by far the most common one brought to counselors of all kinds—philosophy would be the most direct route to healing. Some people are simply not very philosophical, so they'd do better with another type of counselor. Most people can benefit from psychological insight, but understanding doesn't stop there. How will you know the right thing to do if you don't know yourself? Part of knowing yourself is psychological, of course, as well as physical. But ultimately, discovering the deepest essence of yourself is a philosophical task.

If you are chronically depressed from some kind of unresolved trauma in your past, medication may help you feel able to talk about it and so may be useful over the short term. In a very few cases the same might apply if you are dealing with a more immediate crisis. But taking medication then just puts off the inevitable, and the risk is that feeling better because of a pill means you won't do the work that is necessary to meet and surpass whatever challenge lies ahead. Drugs don't do anything in the outside world—even with a mood softened by Prozac, you'll still have to deal with a sadistic boss or a cheating partner or a bureaucratic bank. The answers are not—and never will be—in a pill bottle. The best you'll get there is temporary palliation.

Just as medication might help in more purely psychological or philosophical cases, philosophy can provide additional help in just about any case where physical or psychological work is being done. Even in the strictest psychiatric case, like needing lithium for manic-depression, philosophy could be helpful once you are medically stable. Bearing a diagnosis like that might be easier if you could develop a functional philosophical disposition toward your situation. One of the reasons so many patients have trouble staying on medication—even when they find the medication helpful—is that they somehow don't feel like themselves while taking it. That gets to the heart of the most basic philosophical question there is: "Who am I?" You may need to rediscover yourself on medication. That may in turn lead to the kinds of questions ("What makes me me?" "What am I—if anything—aside from my physical body?") that are philosophers' bread and butter.

EMPATHY, NOT EXPERTISE

A good therapist of any stripe will provide sympathy, empathy, and moral support, which can go a long way toward healing. Something as simple as dialogue with another caring individual is a balm in many cases. It isn't expertise that makes a good counselor; expertise isn't even necessary. More important is the ability to listen, to empathize, to understand what another person is saying, to offer some new way of looking at it, and to proffer solutions or hope. A large part of what you respond to in therapy is the style of the therapist. Someone you personally relate to, whose insights appeal to you, and who sets an example that is meaningful to you will be the person you make progress with in therapy.

Most talk therapies work because of the therapist and the match between the therapist and the client, not the particular school of therapy. Regardless of who the other person is or what he tells you, interaction alone can be helpful. But it isn't an instant cure. There is no immediate treatment for an emotional toothache; no obvious way to fill the cavity or extract the tooth. You must aim to understand your problem, learn to live with it, and move on. Psychological counseling is a way of exploring and coming to terms with your emotional responses to a problem. That's a good place to start. Philosophical counseling is a way of exploring and coming to terms with your problem itself. That's a good place to finish. The latter approach is obviously more direct, focusing on coping with whatever problem you face, finding and taking whatever necessary actions are consistent with your personal philosophy, and using what you learn as you move forward with your life.

For most of my clients, philosophical counseling is a short-term proposition. For most people, psychological counseling is longer term. (The interference of managed care is changing that, though shortening the duration of a relationship designed to be long-term is a detriment to rather than an improvement in the process.) The virtue of therapy itself may mean that almost any therapy is better than no therapy (although the wrong kind of therapy can be worse than no therapy) and that a good psychologist is a better choice for a counselor than a bad philosopher. Sometimes the best option would be to talk to a wise person, no matter what their background. More than a few of us received good advice from our grandparents, who knew and understood a lot about people just by having lived among them for so long. The balance of psychological and philosophical insight is what will actually benefit most people.

Many good psychologists are very philosophical. And the best philosophers are also psychological. Psychotherapy comes in countless flavors, and despite its current medical connotation, I remind you that *psychotherapy* comes from two Greek words that have nothing to do with medicine: *therapeuein* means "to attend to" something, while *psukhé* means "soul" or "breath" or "character." Psychotherapy, then, can mean attending to your soul, which makes your priest, minister, or rabbi a psychotherapist. It can also mean attending to your breath, which makes your yoga instructor, flute teacher, or meditation master a psychotherapist. It can also mean attending to your character, which makes your philosophical counselor a psychotherapist.

The idea that every personal problem is a mental illness is practically a mental illness itself. It is caused primarily by thoughtlessness, and cured primarily by thoughtfulness. And that's where philosophy comes in.

3

PEACE in Your Time:
Five Steps to Managing Problems
Philosophically

---

*"Empty is the argument of the philosopher which does not
relieve any human suffering."*
—EPICURUS

*"The philosophical problem is an awareness of disorder in our
concepts, and can be solved by ordering them."*
—LUDWIG WITTGENSTEIN

Philosophical counseling is more art than science and proceeds in a
unique fashion with each individual. Just as psychological therapy
comes in countless different forms, philosophical counseling has at
least as many permutations as there are practitioners. You can work
through a problem philosophically on your own or with the help of a
nonprofessional partner. The big question is "How?" Some philosophical counselors—notably Gerd Achenbach—claim justifiably that there
is no general method that can be explained or taught. After all, if there's
no general method for doing philosophy, how can there be one for
doing philosophical counseling?

Even so, I have found through experience that many cases conform
successfully to a five-step approach I call the PEACE process. This
approach gets good results, is easy to follow, and also illustrates what
sets philosophical counseling apart from other forms of talk therapy. As
you will see, most of the problems presented in this book were resolved

by the PEACE process. Maybe yours can be handled that way too. PEACE is an acronym that stands for the five stages you go through: problem, emotion, analysis, contemplation, and equilibrium. The acronym is fitting since these steps are the surest path to lasting peace of mind.

The first two steps frame your issue, and most people pass through these stages naturally. They don't need anyone to identify the problem with or for them, though sometimes it is a point to be revisited and refined. Their emotional reaction is immediate and clear—no one needs to learn to feel emotion—though it too may be food for further thought. The next two steps progressively examine the problem, and though many people can do this on their own, it may be helpful to have a partner or guide for exploring new territory. The third step itself takes you beyond most psychology and psychiatry, and the fourth puts you squarely into the philosophical realm. The final stage incorporates into your life what you've learned at each of the first four stages, since solely cerebral approaches aren't practical unless you know how to use them.

I'll give you a brief explanation of each step to show you how the process works. Then I'll go back and explain them in more detail and conclude with a sample case so that you can see the process in action. Each chapter in Part II also describes at least one case in terms of the PEACE process.

When confronting an issue philosophically, you must first identify the *problem*. For instance, your parent is dying, or you've been downsized, or your spouse is cheating on you. We usually know when we have a problem, and most of us have an internal warning system that goes off when we need help or additional resources. Sometimes specifying the problem is more complicated than it seems, so this step may require some work if the parameters of what you're dealing with are not obvious.

Second, you must take stock of the *emotions* provoked by the problem. This is an internal accounting. You must experience emotions genuinely and channel them constructively. Most psychology and psychiatry never progress beyond this stage, which is why their benefits are limited. Following the examples above, your emotions are probably some combination of grief, anger, and sadness, though you might have to do a bit of work to arrive at this conclusion.

In the third step, *analysis,* you enumerate and evaluate your options for solving the problem. An ideal solution would settle both the external issues (the problem) and the internal ones (the emotions stirred up by the problem), but an ideal solution isn't necessarily realizable. To continue with one example, giving the order to pull the plug on your dying mother's respirator might be what is best for her but also what is hardest for you. You might leave the decision up to the doctors, or let your sibling decide, or plan to continue the futile life support—these are the various paths you must walk in your mind to find the most appropriate one.

In the fourth stage, you take a step back, gain some perspective, and *contemplate* your entire situation. To this point, you've compartmentalized each of the stages in order to get a handle on them. But now you exercise your whole brain to integrate them. Rather than dwelling on the individual trees, you examine the shape of the forest. That is, you cultivate a unified philosophical view of your situation as a whole: the problem as you confront it, your emotional reaction to it, and your analyzed options within it. At this point you are ready to consider philosophical insights, systems, and methods for managing your situation in its entirety. Different philosophies offer contrasting interpretations of your situation as well as divergent prescriptions for what—if anything—to do about it. In our example of facing your mother's death, you need to consider your own ideas about quality of life, what responsibilities you have for others, the ethics of withdrawing life support, and the relative weights of competing values. You need to establish, through contemplation, a philosophical position that is both justifiable on its merits and consonant with your own nature.

Finally, after articulating the problem, expressing your emotions, analyzing your options, and contemplating a philosophical position, you reach *equilibrium.* You understand the essence of your problem and are ready to take appropriate and justifiable action. You feel on an even keel but are prepared for the inevitable changes ahead. For example, if you decided to turn off your mother's respirator, you feel sure that it is what she would have wanted and that even if her death is hard on you, it is your responsibility to follow her wishes as best you can in a difficult situation.

EXPLORING PEACE

Some people can work through all five dimensions in a single counseling session; for others, PEACE can take weeks or months. The length of time depends on both the client and the situation. Many clients have already taken themselves through the first three stages—identifying the problem, expressing emotions, analyzing options—before they come for philosophical counseling. If so, the process continues at the contemplative stage. You should proceed at your own pace whether you are working on your own, with a friend or partner, or with a trained practitioner.

Each of us is centered in our own being, looking out at the world from a unique vantage point. We can perceive existence as just a string of things that happen to us and around us, or we can assume some responsibility for much of what occurs. It is part of human nature to think the first way about anything bad and the second way about anything good. When tragedy strikes, you're bound to muse "Why me?" at some point. But that's never the question asked by someone who just won the lottery. If our kids behave and excel, we congratulate ourselves for good parenting. If they act out and rebel, we blame them. Taking responsibility for the good things but dissociating ourselves from the bad is one way to protect and preserve our own interests, and it goes to show that Hobbes was onto something in insisting that people are basically "self-regarding."

For the purposes of defining the problem you face, try to look at what is going on without making judgments about it. You're looking at what philosophers call "phenomena"—that is, events external to you, facts that exist independent of your beliefs, feelings, or desires about them. You could think of this as the phenomenal stage if you're feeling particularly philosophical. As the *I Ching* teaches, things are changing all the time, so we are always encountering new situations.

Fortunately for us, we handle most things routinely. We don't have to scrutinize each new setup, because we have social conventions and personal habits to guide us on most paths. So when you examine your situation, you have to determine what is simply a phenomenon and what is actually a problem for you. Assume—for now—that you are not causing the current situation. You're just living your life, minding your own business. (Later, in the third and fourth steps of this process, you should examine how much of it you are responsible for so that you can take control of whatever part of it is up to you.) You may be swimming in it, but you are not the ocean.

Every time you come across something out of the ordinary, which you have no stock reaction to, you have an emotional response. The limbic system, the oldest part of the brain, generates the physiology of emotion: automatic (technically, autonomic) responses to stimuli. But the experience of an emotion occurs in a higher part of the brain, where your physiological responses are interpreted and labeled. This is a one-way street. The separation guarantees that you can't control an emotion just by recognizing it, a point overlooked by many psychologists and psychiatrists who focus their work on doing exactly that. Understanding that you feel angry isn't going to change your body's angry reaction (e.g., increased heart rate, secretion of adrenaline). Recognizing the emotion is valuable information—just don't expect the insight to contain the feeling. Once you've had the feeling, and identified it, the third part of this step is to express it appropriately. Expression also won't stop the feeling, another note for psychologists and psychiatrists, but inappropriate expression of emotion will probably worsen your situation.

With analysis, you start the process of resolving your problem by taking inventory of your options. You might say, "So, I have this problem, which is making me feel unhappy—what can I do about it?" The most common way to generate alternatives is by analogy. If you are reexperiencing something you've already experienced and resolved, you know what to do, or not to do, depending on how you dealt with the earlier circumstances. You could also consider what happened to your best friend, or what you saw at the movies, or what you read in this book. Finding commonalities with other situations—creating an analogy—is a powerful way of understanding your current difficulty. It may not change how you feel about the problem, but it could give you insight into how or why it is happening and help you generate possible reactions.

Psychological therapies progress no further than analysis—if they even get that far. Many don't; they are mired in endlessly "validating" emotions. Psychiatrists tend to discourage reasoned examination of a problem and instead focus on emotions, to guide you back to childhood. You could do that kind of work for years and never feel better. Again, many people work through the first three stages of the PEACE process by themselves but stop short of contemplation and equilibrium, without which they can't bring about resolution of their problems.

Which brings us to contemplation, and the integration of all the information you've gathered in the first three steps. Your goal here is to adopt a disposition toward—an attitude toward, or a way of regarding—your overall situation. The dictionary would tell you that *disposition* means a prevailing tendency, inclination, mood, or temperament. When you're told you have a cheerful disposition, that's what your admirer is getting at. But in this book, disposition is another way of saying a philosophical outlook. To find yours, you need to take a step back from the immediacy of your problem, the power of your emotion, the logic of your analysis. The crucial step is to adopt an overarching philosophical way of looking at your entire situation that allows you to reconcile yourself and move on.

You might be able to find a philosophy that resonates with you in the work of a recognized philosopher, either by reading the source or by learning the relevant points from a trained philosopher. You surely have a personal philosophy within, although it may not be conscious or articulated enough to work for you. So what you probably need is a guide, or a mirror, to help draw your own philosophy into the open where you can see it and work with it. A disposition is something you find genuinely within. It is more like unearthing a gem than manufacturing a tool. If you claim a philosophical disposition that you don't feel to your core, you will be doing no better than compensating for your situation or rationalizing it. That brings no real or lasting relief. You may also discover that your current disposition is actually making your problem worse and that by changing your disposition you can change your life. That kind of change is beautiful, like the metamorphosis of a pupa to a butterfly. Everything changes, and your key to making the best of change is your disposition.

I sometimes think of this as the cerebral or conceptual stage (more *C*-words). I say cerebral because you're working with your intellect and your emotions: your whole brain. And you need a conception of how everything fits together—all the elements of your situation, all the elements of your world, all the elements of your philosophy. Finding that unity is what allows you to put a problem behind you. If you're stymied by a problem, what you need is a conceptual breakthrough. Your habitual responses aren't enough.

In the final step, you reach equilibrium. With your newly acquired or delineated disposition, you put into action your best option and incorporate what you've learned into your life in a concrete fashion. Your problem ceases to be a problem, and you are returned to your usual but now improved state of being, no longer troubled—until the next time circumstances conspire to throw you out of balance. There is always a certain amount of wobbling; no one remains on an even keel forever. But if you've really made the PEACE process your own, you'll be better equipped for the future. Once you've found a useful disposition, it doesn't go away. You can't exhaust it. You can recall and reuse everything that worked for you in one set of circumstances in any similar situation you face down the line. Whatever works for you gets reinforced, and obversely, what doesn't work gets shelved. If you follow through to this final stage, you won't ever be back at square one. Your life will be enriched, even after the bleakest tragedy, if you can learn about yourself from handling your experience and attaining equilibrium.

I sometimes call this last step essential (another *E*-word), since by the time you get here, you'll have understood the essence of your situation. You'll have discovered not just the essence of your problem, but also something essential about yourself. That's the insight that allows you to resolve your current situation and prepare yourself for the next one. Absolute solutions aren't always possible, so resolution is a more appropriate goal in most cases. This step is also essential because it is the one that allows you to move on. You can become philosophically self-sufficient, so you won't need further counseling (unless you decide to continue beyond the basics). A disposition that leads to equilibrium is something you take with you wherever you go. It doesn't stay in the medicine cabinet to be pulled out when you want to squelch an unpleasant symptom. It isn't something you are dependent on, as you might be on a therapist or a prescription. It is a part of you.

WHAT DO PABLO CASALS AND MARK TWAIN HAVE IN COMMON?

The first two steps of the PEACE process are familiar thanks to my predecessors in psychology and self-help. And as I mentioned, many people find their way through that beginning part of the maze by themselves but then turn to outside help. Since the third and fourth steps— analysis and contemplation—distinguish this method from what has gone before, and since they are the newest and hardest to grasp, I want to illuminate them with two powerful examples. These examples involve famous people who acted as their own philosophical counselors. Later in this chapter we'll look at a case from my own practice.

The great cellist Pablo Casals once broke his arm in a skiing accident and had to wear a cast for six weeks. The problem he faced was clear: having his arm in a cast wreaked havoc on his schedule and interrupted his career. His emotional reaction probably mixed frustration, anxiety, devastation, depression, and fear. His analysis tallied up all the logistical complications: canceling or rescheduling concerts, attending appointments with doctors and physical therapists, calling his agent, rewriting contracts, planning his rehabilitation once the arm was healed, and more.

He held the requisite press conference to share the news with his fans. The gathered reporters might have expected him to be despondent but found instead that he appeared radiant with joy. They asked him why he was so happy. "Because now I don't have to practice," he responded.

I'd be amazed if Casals had consulted a philosophical practitioner, but he clearly grasped the importance of the contemplative stage in facing a problem. He reached inside himself to pull out the most beneficial attitude given his situation. He was obviously aware of the destructive elements of the circumstances but chose instead to focus on the constructive ones. He saw his restrictions as liberating, not limiting. He reframed the issue, asking himself "Now what can I do for the next six weeks that I wouldn't otherwise have been able to do?" I don't know if he took off for a Tahitian vacation, played one-armed bandits in casinos, or spent time with friends, but he surely had more options in that period than he usually would have had, given the demands of his talent. His disposition allowed him to make the most of his situation.

In case you think you don't have that kind of serenity, let me give you another example before you decide this is beyond you. Mark Twain was almost as famous for his fiery temper as he was for his literary achievements. He was easily provoked, and you can rest assured that his ire was withering. When piqued, his rejoinder of choice was to draft a scathing letter. But then he always left the letter on his mantle for three days. If after three days he still felt angry, he mailed it. More often his anger had faded, and he would burn it. That may be a loss for admirers who would covet a copy of those words, but it was surely a boon to Twain and his friends and acquaintances.

I'm betting Twain used those letters to define the problem, express his emotions (primarily anger), and analyze his options (some of which were surely wonderfully graphic). But his contemplative insight was to exercise the virtues of patience, open-mindedness, reflection, and willingness to change. As hot-blooded as Twain was known to be, he could not have shown that kind of restraint unless he was disposed toward those virtues. Whether or not he mailed his angry letter, he used both the letter and the three-day respite to regain his equilibrium.

Though I doubt he was aware of it, Twain was reflecting the Chinese idea that the best course of action is the one that leaves you blameless and with no regrets. By waiting three days and then deciding with a cooler head what to do with the letter, he could feel secure that he would find that path.

Because we are not privy to the internal workings of Casals' or Twain's minds regarding these issues, we don't know with how much ease or difficulty they discovered their dispositions or how much work preceded putting them into practice. Don't make the mistake of assuming it was easy just because the blood, sweat, and tears aren't chronicled here. People willing to make an effort to philosophize about whatever they face can find useful dispositions and attain a measure of equilibrium. It's not necessarily a quick fix. But I can assure you that doing the work and obtaining the result is better than all the alternatives—anger, guilt, escapism, dependency, victimhood, martyrdom, litigation, and the Home Shopping Channel—put together.

VINCENT

Vincent enjoyed a successful career as a professional writer. He had decorated his office space with the usual mementos, photos and the like. He had also put up a print of a famous Gauguin painting, which portrayed seminude Tahitian women on a beach. One of Vincent's female colleagues informed their supervisor that she was offended by the painting and demanded that it be removed. Complying with company guidelines on sexual harassment, the supervisor called Vincent into his office and told him to remove the painting. Vincent objected but had no recourse: he had to take down the art or resign from his job.

After weighing these two options, he chose the lesser of the two evils and decided to remove the art. After all, it's easier to find a new painting than a new job. Vincent did the practical thing. What he hadn't anticipated was the growing anger, outrage, and sense of betrayal of principle he felt after removing the art to keep his job.

As we examine this case through the PEACE process, you will see clearly how philosophical counseling differs from psychological counseling. Many psychologists who listen to my public lectures on philosophical counseling come up to me afterward and say, "You know, I do the same kind of thing as you do." In fact, they don't do the same thing at all, and I often use this case to make my point. I tell psychologists just what I've told you about Vincent and ask them how they'd proceed. Without exception, they focus strictly on his emotions—anger, outrage, betrayal—and say what a lot of work they would do in those areas. In my opinion, that would be a waste of time, not to mention money. When I explain to them how a philosophical counselor (yours truly) proceeded in this case, they suddenly realize that a whole universe of perspective exists unconceived by their psychological training. Here's how it works:

STAGE ONE: *Problem.* Vincent's problem, in a nutshell, was that he suffered from a sense of injustice. He believed he was unjustly compelled to take down the painting and that his job should not have been at stake over a matter of personal taste in art. His emotions flowed from his sense of injustice. That, and not the emotions themselves, was the root problem.

STAGE TWO: *Emotions.* Vincent saw no immediate way to express his emotions constructively. He did not want to feel so angry and so self-betrayed, but the system offered him no remedy for feeling better.

STAGE THREE: *Analysis.* All options considered, Vincent probably did the right thing. He loved his profession, and jobs like his don't grow on trees. If he had resigned over the painting, he'd still feel the same injustice and would now be out of work. It's probably better to be upset and employed than upset and unemployed. If he had a ton of money, he could have sued his employer over the incident and taken his chances in court. But he couldn't afford it. Revenge fantasies were somewhat satisfying in passing but never real options. Anyway, if Vincent had settled on the solution of threatening his supervisor and offended colleague, or going berserk and shooting up the office, he still wouldn't have found justice, but he'd be in jail. All in all, Vincent's choice seemed like the best one he could have made.

STAGE FOUR: *Contemplation.* Philosophically, I worked with Vincent on understanding the distinction between offense and harm. If someone or something harms you—that is, injures you physically against your will—you are not an accomplice to the injury. John Stuart Mill's harm principle states that "the only purpose for which power can be rightfully exercised over any member of a civilized community, against his will, is to prevent harm to others."

But offense is something else. If someone or something offends you—that is, insults you in some way—you are definitely an accomplice to the insult. Why? Because you *took* offense at it. You may be passively harmed by something such as a physical blow, but you take an active part in being offended by something such as a painting. Remember the exchange of bygone days:

"Sorry, no offense intended."

"Well, then, none taken."

That kind of civility has been made obsolete by a culture of careless thinking, which has allowed offense to become confused with harm. Marcus Aurelius knew the difference back in second century Rome, but our advanced culture has forgotten. Nowadays people take offense themselves, then accuse others of harming them, and the system backs this up with policies that undercut individual liberties. Worse, the system reinforces this confusion by rewarding people monetarily for taking offense. No wonder everyone's either walking on eggshells or walking around with their noses out of joint.

> *"Take away your opinion, and there is taken away the complaint*
> *'I have been offended.' Take away the complaint 'I have been*
> *offended,' and the offense is gone."*
> —MARCUS AURELIUS

The distinction between harm and offense was Vincent's first contemplative breakthrough. His second came with the realization that this kind of injustice was systemic and not directed at him personally. His accuser and his supervisor were just pawns in a larger game that they didn't even understand. The absurdity was almost amusing. It wasn't as though Vincent had hung a *Playboy* centerfold on his wall—which some might still consider art, although it's clearly more provocative than a reproduction of a priceless painting. People looking to take offense will always find something to take it at, but then they're the ones with the problem. Their problem is that they need to feel offended. Vincent inadvertently got in the way of someone else's need.

Vincent didn't have to view his situation as unjust, because he himself had been offended, but not harmed, by the system. He had the power to refuse to be offended by the intolerance of the system, and he decided to exercise it. Vincent now had a philosophical disposition that immunized him against this injustice and allowed his negative emotions to dissipate.

STAGE FIVE: *Equilibrium.* Vincent returned to work, no longer harboring ill feelings toward his colleague or his supervisor. He had better things to do than invest emotion in their taste in art; he had a career to get on with. As a final touch, I suggested that Vincent put together a top-ten list of paintings that he would like to hang on his wall, show them to his colleague, and ask her to choose one that didn't offend her. That way, everyone would be satisfied with his decor.

Vincent's PEACE process was accomplished in a single session. At no time did we discuss his childhood, his sexual fantasies, his dreams, his Oedipus complex, or a prescription to improve his mood. The moral: Psychology and psychiatry have nothing to say about injustice. If you want to resolve a philosophical problem, seek philosophical help.

BE YOUR OWN
PHILOSOPHICAL COUNSELOR

Practicing philosophy means exploring your inner universe. You are the one most qualified to undertake this journey of self-discovery, though sometimes you can benefit from the guidance of philosophers who have trodden similar paths. Philosophers almost always work alone in the sense that humans tend to think most clearly in solitude. Yet philosophers almost never work alone in the sense that our thoughts are informed by significant insights from 2,500 years of diverse philosophical traditions. Philosophical counselors are like matchmakers: we help our clients find a philosophical interpretation of themselves and their situations that they can live with, and prosper with, for a lifetime.

You can help yourself philosophically whether or not you know Aristotle from Zen. Follow the steps of the PEACE process. Take what is relevant to you from the next chapter and from the many important philosophical ideas illustrated in Part II and highlighted in Appendix A (Hit Parade of Philosophers). The case studies throughout the book provide examples of how to illuminate your personal concerns with the wisdom of the ages. So armed, you will be able to reach equilibrium on your own in many situations. Keep in mind that not every problem has a quick fix. Great explorations sometimes take more time and effort.

If you are stuck in any particular step of this process, you may need help getting to the next stage. Some people get stuck in the first step, unable to readily identify the nature of the problem confronting them. It's more common to get stuck in the emotional stage, doing things intended to assuage emotions that actually inflame them (like those who drink to escape their problem—their problem being that they drink too much). One can easily get stuck at the third step too, end-lessly analyzing a situation that can't be changed by analysis alone. For those who make it through the first three steps, contemplation can be a challenge. Finding and embracing the right disposition can take min-utes or months—or occasionally even years. But when it brings you to equilibrium, it will have been worth the effort.

If you do get stuck and can't unstick yourself, you may want to consult a philosophical counselor (see the directory in Appendix C). Or try working with another layperson who is philosophically inclined. Use each other as sounding boards to hash out your philosophical outlooks. Sometimes someone has to help you a little so that you can help yourself a lot.

4

What You Missed in Philosophy 101 That Can Help You Now

"Ancient philosophy proposed to mankind an art of living. By contrast, modern philosophy appears above all as the construction of a technical jargon reserved for specialists."
—PIERRE HADOT

"If there is such a thing at all as instruction in philosophy, it can only be instruction in doing one's own thinking."
—LEONARD NELSON

Alfred North Whitehead wrote, "The safest general characterization of the European philosophical tradition is that it consists of a series of footnotes to Plato." In truth, many, many trees have given their lives so that interested parties could read answers to Plato's arguments, or answers to answers to his arguments, or answers to answers to answers . . . well, you get the idea. In the spirit of Whitehead's pronouncement, but allowing a few more threads to be woven in, this chapter offers extremely brief highlights of some important philosophical ideas—the major schools and thinkers I frequently use in philosophical counseling. I hope you'll begin to see how some philosophical ideas can be of direct use in everyday life, whether or not you can spell Maimonides or pronounce Nietzsche. Philosophy, despite its reputation for obscurity and difficulty, can work in practical ways for anyone.

For the purpose of providing a general plan, I've placed the philosophers I introduce here into categories. This is by no means the only way to organize them, so don't be surprised if you see them labeled differently elsewhere. One thing philosophers love is an argument, especially over categories. Even the time periods I attach to the categories are general, with some of the main thinkers of each tradition falling earlier or later. It should be obvious that this chapter is not a definitive look at philosophy's history or all important philosophers. I just want to give you the bare bones of the subject, so you'll have some context for the application of these ideas as they surface in Part II.

Now you're forewarned, so don't sue me over the D on your term paper if you are taking Philosophy 101 and rely on this chapter for all your information. Think of it more as a crib sheet for a cocktail party. If you aren't interested in seeing the big picture right now, and you have no intellectual cocktail parties on your calendar, go ahead and skip to Part II, or to whichever chapter in Part II particularly interests you. You can always revisit this chapter later if your curiosity is piqued. If you want more in-depth information, there are plenty of books devoted to philosophy and its practice, some of which are listed in Appendix D.

I've found it invaluable to cast my net wider than just Western philosophy, but I find myself coming back to a small subset of ideas over and over again. Three important branches of philosophy originated in roughly the same time period of antiquity, circa 600–400 B.C.E. The one responsible for the vision of philosophy conjured in most people's heads—bearded, toga-and-sandals-wearing men—is the Athenian school, featuring Socrates, Plato, and Aristotle. They were building on some significant pre-Socratics too (such as the Cynics and early Stoics), but for our current purposes we'll stick to the biggest guns. At the same time, in a different part of the planet, the Forest Sages of India, most famously Siddhartha Gautama (Buddha), were adding to the Hindu worldview. Just around the globe, Confucius and Lao Tzu were developing Confucianism and Taoism, which, together with the older *I Ching*, form the heart of Chinese philosophy. This crucial period in these ancient civilizations was formative in the history of philosophy.

I use these three traditions with my clients in approximately equal measure, tailoring my choices to the individual, naturally. In Western thought, I find useful strands in philosophers ancient, contemporary, and everything in between. My familiarity with Eastern philosophy is primarily centered in ancient texts whose theory and practice has become widely known and studied in the West, like the *Bhagavad Gita* and the teachings of Buddha. Contemporary gurus of Eastern wisdom tend to earn their followings not so much by setting down new lines of thought as by living in harmony with the wisdom of the ancients. They inspire by example and explication more than by extension. Even so, the writings they draw from are vast and mostly untranslated into English. Some of the Hindu and Buddhist texts I use are sacred to their followers (though I employ them as a source of secular wisdom) and so less open to questioning and rethinking. Judaeo-Christian resources, from the Book of Ecclesiastes to the Beatitudes, also contain useful philosophical insights. So do the works of poets, playwrights, and novelists. Occasionally, so do the utterances of Casey Stengel. Philosophers should not be snobs; we should be grateful to find wisdom where we can.

THE EAST

Indian philosophies—Hinduism, and especially one of its two unorthodox branches, Buddhism—emphasize the cyclical nature of existence, the impermanence of things, the intoxicating effects of desires, and the importance of nonattachment. Attachments, whether to self, others, or things, are the main cause of suffering. One way to reduce suffering, then, is to let go of attachment. Indian philosophy in general—whether Hindu or Buddhist—holds that we should do all things with a whole heart, as a service, not just to reap the fruits that may result from our labor.

Buddhism upholds the moral equality of people but advocates personal responsibility as well as compassion for others. It teaches that one's ruminating mind, grasping ego, and sensual cravings all interfere continuously with the realization of one's lucid serenity (Buddha-nature), and it offers many practices that gradually still the noisy mind, break the fetters of desire, and allow one to remain fully and clearly at peace. One aim of Buddhism is a life free from trouble.

Buddha set down Four Noble Truths (or rather, his scribes and students did; Buddha has left no direct writings). They are strong philosophical medicine, so I don't usually talk to clients about them unless they have suffered enough to be attentive. But these Truths lay out an important path through life's harshest trials, so they can be useful to people in extremely trying circumstances. The first Truth is that life involves suffering. Second, suffering is caused; it doesn't happen by accident. Third, we can discover the cause and break the causal chain to prevent suffering. Remove the cause, and you remove the effect. Fourth, and most importantly, we must practice to achieve the end explained in the third point.

Everything we do has consequences in Buddhist thought, including our moral behavior, although we cannot tell how much time it will take for a consequence to manifest itself or what form it will take when it does. We may not have a choice about being in a particular situation, but we have choices about what we do with the situation we find ourselves in. We choose between good and evil, and if we make good choices, good things will happen. If we make bad choices, bad things will happen. This cedes a measure of responsibility and control to people.

In contrast, Hinduism can result in passivity because of the belief in reincarnation. If this life in its entirety is what you are stuck with as a reward or punishment for a previous life, what difference could your actions make?

Buddhism looks at existence as a series of instants (as opposed to Hinduism's series of lives), and what happens in each instant influences what happens in the next. That's the more optimistic stance I prefer, and the one that demands more personal responsibility. Either way, the thrust is on moral development in contrast to the West's fixation on scientific progress alone or above all.

"I will reveal this knowledge unto thee, and how it may be realized; which, once accomplished, there remains nothing else worth having in this life."
—BHAGAVAD GITA

Chinese philosophy has as a central tenet that everything changes. You shouldn't expect a permanent state of affairs in any aspect of life, and to avoid being mystified by an entirely new situation with each change, you should strive to understand the nature of change. With some insight into why and how things change, change will seem more natural to you, and you will be able to anticipate and do the right thing in moments of change. Philosophers like Lao Tzu (author of the *Tao Te Ching*), Confucius, and the anonymous author(s) of the *I Ching* teach us to make the best of situations we can make the best of—as well as those that aren't under our control, and bad situations too. In every case, we are accountable for our decisions. Despite constant change, the world is seen as an orderly place. To understand the way the human world works, we should understand the way the natural world works and realize the similarities. The common translation of *Tao* is "the Way," signifying the way things unfold. The best way for humans to live is in harmony with natural laws that shape social and political processes.

Chinese philosophy centers on the quest for how to lead a good life. If individuals lead good lives, society will also be good: unconflicted, decent, productive. It does not consider knowledge alone to be the path to the good life (as does much Western thought). Quality of life flows from reflection on duty and morality, interpretation of experience, and understanding of process.

"One cannot easily disregard such great minds as Confucius and Lao Tzu, if one is at all able to appreciate the quality of the thoughts they represent; much less can one overlook the fact that the I Ching was their main source of inspiration. . . . I am now in my eighth decade, and the changing opinions of men scarcely impress me any more; the

thoughts of the old masters are of greater value to me than the
philosophical prejudices of the Western Mind."
—CARL JUNG

THE WEST

In the West, it seems that we philosophers do the philosophical exploration so the rest of you don't have to bother. Returning the importance of personal philosophical introspection to ordinary people is the driving force of philosophical counseling.

Socrates provides one of the more important role models for philosophical counseling, as well as being the "Godfather" of Western philosophy in general. He was Plato's mentor and teacher, and his work survives only through Plato's writings. Philosophical counselors look to him in part because he believed that we all have knowledge already, that what we need to know is within but that we may need help in bringing it out. If you're grappling with a major issue, then what you may need is a sort of philosophical midwife to work with you to elicit your own wisdom. Unlike doctors and lawyers, whose help you seek because they have specialized knowledge that you don't have, philosophical counselors don't necessarily rely on their particular expertise but on their general ability to conduct an inquiry. We don't give you answers but help you ask the profitable questions. We don't necessarily act as authorities revealing information you never knew but provide the guidance many people need, having forgotten or neglected the means of examining themselves.

The other major contribution Socrates left to philosophical counselors (along with countless others who find it useful) is the so-called Socratic method: asking a series of questions to get at ultimate answers. Socrates' famous dictum "The unexamined life is not worth living" sums up his belief that leading a life of quality is the most important thing and that we do so primarily through inquiry.

Plato—the cause of all those philosophical footnotes—was an essentialist. He believed in the existence of pure but abstract Forms of which material objects are imperfect copies. These Forms are changeless, but we live in a world of shifting appearances and can access them only with our minds. With the proper education, we can break through to the noetic world (the world of ideas) and understand the pure Forms of Justice, Beauty, and Truth—and so make better copies of them in the real world we live in. Plato thought the highest calling was to seek the essences of things this way. This idea resonates for philosophical counseling, too, because if you didn't know the essence of something, how would you recognize whether you had it? For example, what is happiness? Fulfillment? Morality? Honing your understanding of concepts like these gives you a philosophical perspective on where you stand in your own life.

Aristotle developed the importance and uses of critical thinking, setting the stage for centuries of philosophical inquiry. He pioneered many physical and social sciences, but little of that is scientifically useful today since he was all theory and no practice—he performed no experiments and wasn't particularly concerned about evidence. He also invented logic, which in its elementary form is very helpful to clients whose problems involve mistakes in critical thinking.

Aristotle's theory of ethics is also extremely important. He defined goodness as the virtue all reasonable creatures strive for, an optimistic view if ever there was one. Virtue depends on a person having a choice, and Aristotle believed the virtuous choice to be the Golden Mean, or the happy medium between extremes. Courage, for example, is the Golden Mean lying between rashness and cowardice. He thought happiness would follow from virtue and goodness and emphasized duty, obligation, and character development as important human concerns. He placed high moral value on temperance and moderation, as you might guess about the man who gave us the Golden Mean. His thoughts about all these components of the good life are very useful in philosophical counseling when a client is working out his or her own conceptions.

"Our present study is not, like other studies, purely theoretical in intention; for the object of our inquiry is not to know what virtue is but how to become good, and that is the sole benefit of it. We must, therefore, consider the right way of performing actions. . . ."
—ARISTOTLE

THE ROMAN EMPIRE AND
THE ROMAN CATHOLIC CHURCH

After Aristotle, things stagnated in the West—from the point of view of what is useful to me in philosophical counseling—for a long time. The tremendous amount of territory that had been charted during the "Great Leap Forward" of Hellenic culture (including philosophy) remained uncultivated for centuries. The military might of the Roman Empire retarded philosophical evolution for a long time—with notable exceptions like the Roman Stoics—possibly because so much energy was devoted to conquest.

After the Roman Empire declined, the political and spiritual grip of the Roman Catholic Church thoroughly controlled European thought, and the only scholarship permitted was strictly religious. The vast majority of people did not know how to read, and most of what few texts existed were in Greek, Latin, Arabic, or Hebrew. Laypeople heard the official interpretation of the scriptures that ruled their lives but did not have access to them. No dissent was allowed. Many books were banned. Many people, including philosophers, were burned at the stake. Europeans had much faith but did little inquiry. Without free thought, nontheological philosophy ground to a halt.

The Roman Catholic Church wielded such enormous power that in 1651 Thomas Hobbes called it "the Ghost of the deceased Roman Empire, sitting crowned upon the grave thereof." In a sense the Church had more power than the Empire. The pen is famously mightier than the sword, and the power of ideas—of doctrine—is longer-lived than the authority of mere governments. Even the greatest empires, relying on the power of the sword, cannot last forever. Spiritual and ideational powers are stronger in the long run. But under the Church, the human capacity for reflection and skepticism were severely restricted and dogmas were accepted without question. Philosophy, on the other hand, questions everything. This fundamental dichotomy between theology, which requires faith, and philosophy, which exercises doubt, often makes the two fields incompatible, as they certainly were for more than a millennium, until the Reformation and the eventual beginning of the Scientific Revolution.

I am not singling out the Roman Catholic Church—all religions function in this way. Every religion has core beliefs that are supposedly unchallengeable, until some philosopher comes along to challenge them. The Roman Catholic Church complicated this issue because, thanks to Aquinas, its theology also embodied Aristotle's metaphysics and science, much of which turned out to be nonsense. But you couldn't say that about Aristotle's philosophy without being accused of heresy against Rome. That's what got Galileo into such hot water (and almost into a hotter fire) in the seventeenth century: by proving that some of Aristotle's physics and astronomy were dead wrong, he was inadvertently saying that associated Roman Catholic doctrines were dead wrong, which was a capital crime in his day.

But religions evolve too. The Roman Catholic Church was famous (or infamous) for its Index of Prohibited Books. Hobbes's *Leviathan* was banned as soon as it appeared, but it was neither the first nor the last great book to run afoul of religious politics. In the twentieth century, the Index has at one time or another banned the *Complete Works of Freud*, as well as books by Aldous Huxley, James Joyce, Alfred Kinsey, Thomas Mann, Margaret Mead, Bertrand Russell, H. G. Wells, and others. One might ask, "How can human beings ever mature beyond Freudian self-conceptions if they're not even allowed to entertain Freudian self-conceptions?" But times are rapidly changing. Under Pope John Paul II, the Roman Church now recognizes that Darwin's *On the Origin of Species* (number one on the Index since it appeared in 1859) is compatible with Genesis. If you think that means all bets are off, you're right. More recently, Pope John Paul II's encyclical *Fides et Ratio* (*Faith and Reason*) exhorts all Catholics to focus on philosophy. "The pope has a cast of philosophical heroes that would have made former pontiffs blench," reports the *London Daily Telegraph*. He admires not only Western philosophers but also India's sacred texts, the teachings of Buddha, and the works of Confucius. So now we inherit a new rhetorical question: instead of asking, "Is the Pope Catholic?" we can ask, "Is the Pope philosophical?"

> "... *many people stumble through life to the very edge of the abyss without knowing where they are going. At times, this happens because those whose vocation it is to give cultural expression to their thinking no longer look to the truth, preferring quick success to the toil of patient inquiry into what makes life worth living.*"
> —POPE JOHN PAUL II

This new partnership between theology and philosophy extends right to the grass roots. On a personal level, my frequent companion at Zen and breakfast is the esteemed Roshi Robert Kennedy, SJ. That's right; he's both a Zen master and a Jesuit. (His book is recommended in Appendix D.) Still more: the first American institution of accredited higher learning to offer my graduate-level course on philosophical practice is Felician College, a small Catholic college in New Jersey. From my perspective, the Roman Catholic Church is promoting a far-reaching philosophical renaissance. How interesting and wonderful for Catholicism and philosophy both.

THE EARLY MODERNS

The early modern philosophers, who emerged in the seventeenth century, marked the passing of the Dark Ages. After the philosophical revolution fomented by Francis Bacon, Thomas Hobbes, René Descartes, and Galileo among others, the world could never be the same again. By declaring, "Knowledge is power," Bacon provided a crucial third way between faith and the sword: science. The word *science* hadn't even been coined at that point, but Bacon's focus on empirical knowledge laid the foundation for a new way of experiencing—and experimenting on—the world. He emphasized the importance of generalizing from specific instances of physical phenomena to hypotheses that could then be tested. If power came from knowledge, knowledge came from experimentation. He held that both experience and reason are necessary to know the world. The world owes a debt of gratitude to Bacon for giving us the scientific method.

> *"Human knowledge and human power meet in one, for where the cause is not known the effect cannot be produced. Nature to be commanded must be obeyed. . . . The subtlety of nature is greater many times over than the subtlety of the senses and understanding."*
> —FRANCIS BACON

Bacon came from a financially and culturally privileged family, and he employed a raft of secretaries to take down his orated wisdom—in Latin. As a young man, Thomas Hobbes worked as one of Bacon's scribes. And he learned well, later surpassing his mentor. Another of the great early modern thinkers, Hobbes was the first political scientist and the first empirical psychologist. He set out to learn about human nature by making observations about what humans do, with no particular theory at the outset (a method Freud would later make hay with). Hobbes observed that humans are by nature self-regarding and need the controlling influence of civilization and authority to keep the peace.

We can thank René Descartes for giving us a famous articulation of the coexistence of mind and matter. The acknowledgment of the dichotomy between mind and body, and their complex interrelationship, makes philosophical counseling possible. Your mind, as distinct from your brain, may have questions, doubts, false information, shaky interpretations, and inconsistencies—but it cannot have illness. Illness is a physical problem, so if you do have a problem that originates in your brain—or suspect you do—you should see a doctor. Ideas and beliefs are plausibly mental states, not solely or not at all physical ones; hence a philosophical counselor relies on Descartes's distinction to make a place for cultivating minds as opposed to treating bodies.

Descartes's other contribution is more practical than the theoretical one above. As the king of all skeptics, he set himself the mission "to never accept a thing as true until I knew it as such without a single doubt." He believed in carefully examining all we know (or think we know) to separate true beliefs from uncertain or false ones. He held up everything to a critical light to see if it could withstand scrutiny and refused to accept even the smallest piece of information imparted through teachers or through his senses. His other famous contribution, "I think, therefore I am," shifts the responsibility for discovering what is true and right to the first person. That concept also lays groundwork for philosophical counseling by placing a premium on thought (not the senses and not emotion) as the key to understanding.

Galileo had the courage to look into the nature of physical phenomena and report what his observations revealed—even when they contradicted accepted doctrine. He stood by the premise that if the facts don't conform to the theory, then the theory, not the facts, is wrong. For example, Aristotle had declared the moon to be a perfect sphere, but Galileo, the first person to turn a telescope to the heavens, immediately observed that the moon had craters and mountains. He was accused of heresy! Had psychiatry existed in the seventeenth century, Galileo's case might have been diagnosed as noncompliance with Aristotelian perfection disorder, or lunar crater and mountain syndrome.

THE EMPIRICISTS

These early modern philosophers paved the way for the British empiricists Hume, Berkeley, and Locke, who agreed that perception and experience were key to understanding the world. Plato believed that we know what we know through reason and that we are basically born with that knowledge inside of us, though we may need a guide or midwife to get it out. The empiricists were working in direct reaction to this idea, putting the focus on experience rather than reason. They contemplated what it means to experience things through our senses. Like the early moderns before them, they helped set the stage for more rigorous scientific inquiry.

John Locke's famous legacy is the idea of the mind of a newborn child as a tabula rasa, or blank slate. He believed that our minds are completely impressionable and that all knowledge is impressed upon us from outside ourselves. He divided ideas acquired from experience into two types: sensations—the information we get through seeing, hearing, and our other senses—and reflections—the information we get through introspection and processes of the mind like thinking, believing, imagining, and willing. Though I think he is only partly right, his ideas are powerful for philosophical counseling. Many differ from Locke, for example, in thinking that we have an innate ability to learn language, while Locke argued that there are no innate ideas. But where he's right is that the language we do learn is the one we hear—the one we experience.

Starting with a blank slate means that children acquire values and prejudices long before they have the ability to form their own. And children accept many things uncritically. This has important implications for parents, teachers, and anyone else responsible for molding young minds, and explains the tremendous responsibility that comes with that opportunity. It also suggests that if our children are committing suicide and homicide, abusing drugs, and running amok with ever-increasing frequency—which they are—then there is something drastically wrong with the lessons they are being taught. It also suggests the potential of philosophical counseling to help find the eraser if what has been written on your blank slate is misleading or harmful to you, and to help chalk up new ideas that are more pertinent or more helpful.

David Hume took empiricism to an extreme, believing, like Locke, that we have no ideas outside of our experiences and that "all our ideas are copied from our [sense] impressions." He also thought that there was no such thing as necessary cause; that we can't establish a causal connection between any two events. We frequently assume connections (if I hit this ball with a stick, it will move) based on our past experiences, but there is no guarantee that just because something happened in a certain way in the past, it will continue to happen that way in the future. And just because one thing regularly follows another is not enough to prove that the first caused the second. As difficult as it can be to wrap your mind around, this argument can be very liberating. Denying necessary cause is the same as saying that there is no predestination, no fate. This is a key that opens the door to the belief that you can change.

THE RATIONALISTS

The eighteenth century rationalists, headed by Immanuel Kant, extended the footnotes to Plato by reemphasizing reason. But where Locke would say that experience is the only chalk that marks the blank slate, Kant and the rationalists align with Plato by contending that reason leaves impressions there too. Where the empiricists look at trying something to see if it works and then improving it based on results, rationalists look at how things work in the first place. Rationalists believed that one form of knowledge led to another and that it was possible to find a way to unify all knowledge.

Kant recognized that reason, too, has its limits. In his famous *Critique of Pure Reason* he explained his theory that the world is divided into the phenomenal realm (the stuff we can sense; the world as it appears to us) and the noumenal (the world as it really is). Sounds like . . . another footnote to Plato! Kant held that things are one certain, particular way but that all we can know are appearances. Whether you're looking at atoms, rocks, relationships, or societies, you can observe things in many ways. For example, look out your window at a tree. Now do the same in the middle of the night. Try it on a rainy day. Then use an infrared device to look at it. Imagine what it looks like to a bat, or an elephant, or to someone who is color-blind. What does the tree really look like? One of these ways? None of these ways? Kant would argue for the sum of all possible ways plus all unperceivable ways as being the noumenal way. Thus the "thing in itself"—the thing as it really is—is much richer, deeper, and more complete than any one specific phenomenal representation can be. Our reason can tell us only about the phenomenal world.

> *"Reason does not however teach us anything concerning the thing in itself: it only instructs us as regards its own complete and highest use in the field of possible experience. But this is all that can be reasonably desired in the present case, and with which we have cause to be satisfied."*
> —IMMANUEL KANT

In philosophical counseling, it is important to keep in mind that our current perception is just one way of seeing things and that the more perspectives we can investigate, the better our understanding will become. Kant's work also cautions us against defining categories or making judgments, because it is difficult to know whether the category or judgment reflects the thing or the way we are looking at the thing. Anaïs Nin encapsulated this idea when she wrote, "We don't see things as they are, we see them as we are."

Kant's theory of ethics is also important in philosophical counseling. He belonged to the deontology (rule-based) school of thought, as opposed to the school of teleological, or consequential, ethics, which holds that actions are right or wrong depending on the goodness or badness of the outcome. To teleologists, Robin Hood is a hero because, basically, his ends (giving to the poor) justify his means (stealing from the rich). Deontologists like Kant, on the other hand, believe that a rule is a rule: stealing is wrong. They'd have put Jean Valjean in jail for stealing that loaf of bread, never mind his hungry wife and kids.

The strength of the deontological school is that you have a rule book (be it the Bible, the Koran, the Scout's Handbook, or your own personal manuscript) to consult if you are struggling to find the right path. It is usually easy to agree on the basic rules. The drawback is that with a definite set of rules there will always be exceptions (just as killing in self-defense and capital punishment are widely accepted despite the commandment "Thou shalt not kill"), and it is never easy to agree on the exceptions. On the other hand, with consequential ethics, you never know what is right or wrong until you know the outcome, which makes it hard to plan ahead. Its strength is its flexibility and open-mindedness.

On one level, it is helpful for individuals to identify which kind of ethical system they have and which kind they admire. Kant takes it one step further, adding an unusual rule for a deontologist. He believed that you can and should test your decisions for moral and ethical soundness and outlined a thought experiment he called the Categorical Imperative to help you do just that. When considering any course of action, ask yourself, "Would I want everyone else, if placed in my position, to do the same thing?" If the answer is yes, you're on the right path. If the answer is no, then don't do it yourself. For example, while you can easily imagine a situation in which it might be to your advantage to lie, you would not want everyone to lie, so you should not lie yourself.

"I am never to act otherwise than so that I could also will that my maxim should become a universal law."
—IMMANUEL KANT

THE NEW-OLD SCHOOL

While Europe remained the hotbed of philosophical thought, the young country of America was greatly influenced by what was going on across the Atlantic. Two framers of the American constitution, Benjamin Franklin and Thomas Jefferson, were experimental philosophers themselves. Or tinkerers, to use rough terminology more in the spirit of early America. Inventors, naturalists, collectors of specimens—these founding fathers followed the empiricist tradition. Like their colleagues John Adams and Thomas Paine, they also inherited from the rationalists a taste for the power of reason and from the humanists a celebration of moral equality and individual achievement. I can think of no better Exhibit A than the Constitution. The founding of America owed much to the philosophical vibrancy of its creators—a group of philosophical practitioners the like of which we haven't seen since.

THE ROMANTICS

All this attention to hard-and-fast knowledge prodded the development of the nineteenth century romantic revolution. Empiricism gave us better equipment, rationalism gave us better theories, and the resulting scientific and engineering advances brought the world unprecedented technological progress. But in the Scientific and Industrial Revolutions, the romantics saw the worst facets of empiricism and rationalism. Though all this progress was meant to be in the service of humanity, romantics saw that too often those noble aspirations were lost in a sea of exploitation. With wars becoming ever more ghastly as weapons improved, slavery expanding abroad, and women and young children worked ruthlessly in mines and factories at home, romantics looked around and saw their predecessors' ways making things worse, not better. Romantic philosophy developed in reaction against materialism, the mechanization of society, and the view of people as cogs in a machine.

In contrast, romantics focused on the uniqueness of each individual, the importance of spirituality, and the power of art. They valued nature over civilization and emotion over intellect. Though he was actually an eighteenth century figure, Rousseau is the prototypical romantic. His idea of the noble savage—left in a state of nature, we would be our best selves, but civilization is corrupting—informed much of what followed. In Germany, a different version of romanticism arose, called idealism, pioneered by Hegel (more of him later). In England, romanticism produced more poetry—Byron, Shelley, Keats, Wordsworth, Browning—than philosophy, but the impulses were the same.

> *"To her fair works did Nature link*
> *The human soul that through me ran;*
> *And much it grieved my heart to think*
> *What man has made of man."*
> —WILLIAM WORDSWORTH

I'm not a big fan of Jean-Jacques Rousseau, because many of his ideas are naive—and because he himself didn't live by them. He believed "Man is naturally good, and only by institutions is he made bad." This stands in stark contrast to Hobbes's view of life without government as "solitary, poor, nasty, brutish, and short." I believe the truth lies somewhere in between. Humans are no doubt self-regarding, and if unchecked, that element can reach unpleasant extremes. But we also have elements of good in society at large. We can be generous, fair, and just. Most people are capable of leaning either way, an idea Aristotle and Confucius would have agreed with.

The debate between nature and nurture may never be resolved conclusively, but it is clear that upbringing does play quite elaborate tunes on the strings provided by nature. Americans love this kind of adversarial circus—Rousseau versus Hobbes, Democrats versus Republicans, Mothers Who Nag versus Daughters Who Hate Them—but nothing is ever resolved this way. We think the truth emerges through confrontation, but the result is often confusion. Clashes are good for bringing important ideas to our attention, but the world is not black-and-white.

Hegel's idea of the dialectic helps advance the cause of venturing beyond black-and-white thinking. In conflict, he believed that one should present a thesis and an antithesis, then reconcile them through synthesis. Synthesis requires seeing the truth and falsehood in each point of view in order to arrive at something better than both. Before you think this sounds too simple, let me add that Hegel thought we should then propose the synthesis we arrive at as a new thesis, counter it with a new antithesis, and hash out a new synthesis, ad infinitum, until we reach the ultimate synthesis, the Absolute Idea, or truth. Even if you don't want to continue into infinity, this kind of constant refinement is a useful approach to your personal philosophy of life.

Another of Hegel's legacies to philosophical counseling is the idea of transcendence. For Hegel, *to transcend* means both "to negate" and "to preserve." Your identity is like a series of concentric rings. The innermost one is your personal being, then comes your family, then your community, then your town or city, then your state, then your country, then your planet, and so forth. Your actions promote the exclusivity or inclusivity of each level. If you serve your community, then you also serve your family, hence you have preserved that level's inclusivity. You have transcended your family—yet you have also served them, and yourself too. Similarly, you transcend your city when you serve your country, and transcend your country when you serve humankind.

THE UTILITARIANS

In the nineteenth century, utilitarianism was born out of a conviction that the Industrial Revolution had failed. While the tremendous objective advances of science and technology could not be discounted, utilitarians argued that it had failed nonetheless because it did not improve the quality of life for most people. The souvenir T-shirt of a utilitarian gathering would read "The greatest happiness of the greatest number." That phrase belongs to Jeremy Bentham, the founder of University College London, the first English university to admit women, Jews, Catholics, dissenters, and other "social undesirables" of the day. (I'm proud to say it admitted me too—I completed my graduate studies there.) Its equal opportunity policy was an outgrowth of the utilitarian commitment to social justice. To utilitarians, happiness was not an individual condition, but a state made possible by an equitable and useful social structure. They visualized the ideal society as an oval, with a large and prosperous middle, rather than a pyramid with a tiny group of prosperous members at the top supported by a huge base of the less affluent. Utilitarianism is more than a useful way to think about social policy, though that is mainly what its proponents had in mind. It is also instructive for living in a family, or in a smaller community, or within a group of friends, which is how it comes up in philosophical counseling.

"Nature has placed mankind under the governance of two sovereign masters, pain and pleasure. It is for them alone to point out what we ought to do, as well as to determine what we shall do."
—Jeremy Bentham

John Stuart Mill is the leading light of utilitarianism. He studied with Bentham and ultimately surpassed him with his own contributions. Though it wouldn't fit as neatly on a bumper sticker as Bentham's, Mill's formulation of utility, or the greatest happiness principle, is somewhat more specific: "Actions are right in proportion as they tend to promote happiness, wrong as they tend to produce the reverse of happiness. By happiness is intended pleasure and the absence of pain; by unhappiness, pain and the privation of pleasure." This focus on actions and their consequences—while ignoring motives—is problematic but has been highly influential.

A freethinker and libertarian as well as a utilitarian, Mill was also an egalitarian and published an essay criticizing the subjugation of women. In this he was ahead of his time, although his work did follow Mary Wollstonecraft's *Vindication of the Rights of Women*. Mill was a great defender of individual liberty in general, and one of his most famous works is *On Liberty*, where he puts forth his harm principle. Recall, he thought that the only justification we have for restraining a person is to prevent that person from doing harm to others. He went so far as to say that you even have the right to harm yourself so long as you don't also harm others. Get falling down drunk every night if you want, Mill would say, so long as you don't drive drunk, use your children's lunch money to buy your booze, or beat or neglect your spouse. This has implications for your personal life as well as government, and also influences the ethics of philosophical counseling in an important way. If a client came to me saying he planned to harm someone, I would refuse to counsel him and would probably intervene to try to stop the harm from occurring. Some counselors might well take a different stance in the interests of confidentiality—the doctor-patient privilege. In contrast, I recognize a secondary responsibility to the community at large as well as a primary one to my client.

THE PRAGMATISTS

The one school of modern philosophy that is uniquely American, pragmatism developed as a reaction against the smugness of rationalism and the naïveté of romanticism. Its three founders were Charles Sanders Peirce (pronounced Purse), William James, and John Dewey. Although they naturally differed among themselves on many points (what group of philosophers doesn't?), their central idea was that the truth of a theory, or the rightness of an action, or the value of a practice is demonstrated by its usefulness. The best tool, in other words, is one that gets the job done. This view is quintessentially American: durable, portable, and practical. If something's good for you, it's good. I like to think the original pragmatists would have given two thumbs up to philosophical counseling: it helps people, so it is pragmatically valuable.

"We must find a theory that will work. . . . [O]ur theory must mediate between all previous truths and certain new experiences. It must derange common sense and previous belief as little as possible, and it must lead to some sensible terminus or other that can be verified exactly. To 'work' means both these things. . . ."
—WILLIAM JAMES

THE EXISTENTIALISTS

Existentialism emerged at the end of the nineteenth century when much of intellectual thought was collapsing on itself. Many people had thought humans were on the verge of acquiring all knowledge. The line of reasoning went that there were just a couple of problems left in physics and mathematics and that once they were solved, our knowledge of the theoretical and natural world would be complete. That accomplishment would spill over into the social world, and we'd soon be back in Eden again. The ancient Greeks had a word for this kind of exaggerated self-confidence: *hubris*. And it usually preceded a bigger fall than pride alone.

Of course, just as it seemed we were getting to the bottom of it all, not only did new unanswered questions arise, but new unanswerable questions popped up as well. Einstein's relativity theory showed us that length, mass, and time were not absolutes, but that things are measured in relation to other things—only the speed of light appears invariant. Quantum theory (and the Heisenberg uncertainty principle) showed us that despite sophisticated equipment, the fabric of submicroscopic nature contains pairs of things we cannot measure precisely at a given time. Gödel's incompleteness theorem showed that there are theorems we will never be able to prove or disprove—thus some questions in mathematics will never be answerable. As we came to grips with this sudden loss of absoluteness—that is, were sentenced to imperfect knowledge in logic, mathematics, and physics—we confronted even greater gaps of knowledge in the biological, psychological, and social realms. We could no longer look to the sum total of knowledge to make us wise. Scientific and technological progress had to be tempered by new philosophical insights.

The existentialists stepped right into that gap. They rejected the Platonic essentialism (and the idea of perfect knowledge) that had dominated philosophy to this point. They believed that there was no initial essence, only being. What you see is what you get. If there is no essence, the argument goes, we are all hollow. From that perspective,

Nietzsche declared, "God is dead!" (adding elsewhere, as if that alone weren't enough of a downer, "And we have killed him").

Thinking about a universe of randomness and indifference leads many into the depths of despair. We are deprived of the rich and highly textured fabric that connects us to one another. It is an alienating, isolating, soulless worldview at first blush. The feeling behind all this is "So why get up in the morning?" Søren Kierkegaard—generally considered the first existentialist, even though he had a Christian bent, in stark contrast to the atheism of most existentialists—called the reaction to facing this view of our lives "dread." Sartre called it "nausea": "Everything is gratuitous, this garden, this city and myself. When you suddenly realize it, it makes you feel sick and everything begins to drift . . . that's nausea." Indeed, some regard existentialism as more of a mood than a philosophy, and some of its leading texts are in fact novels (notably by Sartre and Camus) rather than philosophical treatises.

But the key point is often neglected: the existentialists were on a moral quest to do the right thing in the absence of some essential idea of goodness and bereft of divine authority. They argued that we must do the right thing even when there is no reason to do it and that real courage and integrity mean doing the right thing for its own sake. That's a breath of fresh air: doing the right thing not because we fear punishment, or crave accolades, or find it expedient, or wish to avoid sinning—but simply because it is the right thing to do. Bad things, then, happen just because they do, not necessarily as some kind of punishment, freeing us from guilt. We must still recognize right and wrong; in fact, we have more reason than ever to find the ethical way. That's the kernel of hope and goodness at the heart of existentialism, often so cloaked in depressing rhetoric that it's easy to overlook. The existentialists actually rediscovered morality. In their line of thought, it may be all there is.

Kierkegaard realized the difficulty of confronting pure existence—no essence, no mystery, no intangibles, no meaning, no purpose, no value. An abyss looms, where hope, progress, and ideals look like illusions. Your existence becomes very thin, and the easy trap to fall into is to wonder why you are alive at all. Religious beliefs can be very comforting, whether true or not, and when existentialism or anything else knocks them out of the way, anxiety may well follow. I don't use Kierkegaard as much as some of my philosophical counseling colleagues do, but just identifying the source of your angst can be helpful. For people who feel anxious but don't know why, it is worthwhile to sort out whether the anxiety is over a particular circumstance (waiting for medical test results, anticipating life after divorce) or a more abstract existential worry. Many people go through an existential phase and gradually build meaning and purpose back into their lives, eventually leaving the angst behind. If existentialism has got you down, try thinking of it as just a phase, and see what you can do to advance beyond it. Once you are through an existential crisis, you may well feel more at peace. It's one way to get rid of a lot of excess baggage.

Frederick Nietzsche is most remembered for his idea of man and superman. He thought each person had a duty to evolve, to strive to be a superman. One way to look at this is as a call to be your own best self, or to lift yourself up above the common standard. Nietzsche himself had an unhealthy contempt for the average person; he believed that rising above meant rejecting conventional morality, and his ideas were badly abused by the Nazis. To use his work you need to winnow the grains of wisdom from the chaff of venom. But his belief that we are too easily satisfied with mediocrity and that most of us don't bother to be all that we can be is a warning worth heeding.

Jean-Paul Sartre explored another logical extension of existentialism: if the universe is undetermined, we are completely free to choose our own course. While constant possibility—with the responsibility for action always falling on the individual—can be a daunting proposition, it is also liberating. No matter what your past experience, you control your future direction. Sartre labeled as "bad faith" any efforts to deny that we are responsible for our actions, and he saw religion, or religious faith, as one of the leading culprits. In calling existential angst nausea, Sartre also connected the mind and the body on some level, acknowledging that the disorienting effects of existentialism can be physically discomforting.

"Man is nothing else but that which he makes of himself.
That is the first principle of existentialism."
—JEAN-PAUL SARTRE

In the novels of Albert Camus, including *The Plague* and *The Stranger*, as in Sartre's novel *Nausea* and play *No Exit*, the heroes are always trying to do the right thing, even when everything is falling apart. They are good people, although they suffer a lot; they are numb but still strive to do good. Camus, who won the 1957 Nobel Prize for Literature, was particularly interested in the absurd, which he used to describe the feeling of meaningless existence. Whether absurdity, nausea, or dread is the presenting symptom, existential crises are commonly encountered—and resolved—by philosophical counselors.

ANALYTICAL PHILOSOPHY

Analytical philosophy emerged at the same time as existentialism, around the turn of the century. Analysis, in philosophical parlance, means breaking down a concept into the simplest possible parts to reveal its logical structure. This field aimed at explaining things in terms of logical structures and the properties of formal language. I admire the pioneers of this branch of philosophy—Bertrand Russell, Gottlob Frege, Alfred Ayer, and G. E. Moore—for their rigorous focus on logic. But in their rigor they also excluded everything emotional, intangible, essential—in other words, everything else. Though Russell himself wrote more than seventy books touching on every conceivable human subject and involved himself passionately (if not recklessly) in social causes, the school of thought he helped found gradually removed itself from the human world. Philosophy, pursued almost exclusively in the academy from this point, eventually become so insular, specialized, and inscrutable that it bore less and less relevance to daily life and become more and more inaccessible to ordinary people.

Philosophy used to address the physical world and investigate the inner workings of human nature. But science has taken over those areas, so what is left for philosophy to do? My answer is: less or more, depending on your approach. How to think critically and how to lead a good life were central concerns of philosophy from ancient times, all but removed from the public agenda in recent decades. Meanwhile, the tradition of analytical philosophy has continued in at least three main branches: the philosophy of language, the philosophy of science, and the philosophy of mind. Each develops productively, or not so productively, depending on the philosophers themselves.

At its best, the philosophy of language reveals and explores important structures and properties of this marvelous human capacity. At its worst, it insists that our sense of meaning emerges from such structures and properties alone (but fails to show how). At its best, the philosophy of science explains how science works and explores the philosophical assumptions that scientists must make in order to perform experiments and interpret their results. At its worst, when done by philosophers who don't know any science, it degenerates into empty theorizing about ill-understood theories. At its best, the philosophy of mind attempts to understand the differences between the mind and the brain, between the brain and the computer, between computing and consciousness. At its worst, it spends its time arguing that we just might be mistaken in believing that we have beliefs, or that we can't be sure that we're thinking thoughts just because we think we're thinking them. At its best, analytical philosophy remains on the cutting edge of human understanding because philosophers are always on the lookout for human misconception, which abounds in every age. At its worst, analytical philosophy is a harmless pastime—which is still a lot better than most other things at their worst.

In case you think I'm bashing analytical philosophy, you can get all this from the horse's mouth. Willard Quine, probably America's most famous and respected analytical philosopher, is even less diplomatic than I—and I'm not renowned for diplomacy!

"Granted, much literature produced under the head of linguistic philosophy is philosophically inconsequential. Some pieces . . . are simply incompetent; for quality control is spotty in the burgeoning philosophical press."
—WILLARD QUINE

APPLIED ETHICS AND
PHILOSOPHICAL COUNSELING

The wheel of change revolved anew with the emergence of applied ethics in the mid-1980s. Biomedical, business, computing, and environmental ethics are currently the most familiar ways of using philosophical tools to analyze problems in the real world. This still-growing industry puts philosophy to work on some important issues of our time. These problems surface because scientific and technological change forces us to rethink and redraft existing laws—whether about euthanasia, employing illegal immigrants, allowing hate speech on the Internet, or disposing of toxic waste. Before we create or amend legislation, we need to clarify our philosophical positions. There are even textbooks on legal ethics and journalism ethics, but evidently they are not widely read.

Last but not least, during the past twenty years in Europe and the last decade in this country, philosophical counseling has begun to extend the meaning of practical philosophy. Applied ethics reaches across the professions and sometimes tackles subjects of importance to individuals. Some applied ethicists are also counselors and consultants; thus they are philosophical practitioners too. But applied ethics usually addresses large issues more than personal problems.

By contrast, philosophical counselors work with individual clients. Gerd Achenbach lit this fuse in 1981, when he opened his philosophical counseling practice in Germany. The "movement" of philosophical practice is now exploding into public awareness worldwide. My colleagues and I draw on the collective wisdom of the ages to guide our clients in philosophical directions that help them resolve or manage their problems. We deal with all the topics of modern (and postmodern) life that make living so challenging, complex, and ultimately, worthwhile and rewarding. We're helping people lead the examined life.

I hope this particular footnote to Plato can help recast philosophy as a dynamic tradition, not an obscure one. We practitioners seek to build bridges between the accumulated wisdom of the past two and a half millennia and the copious need for fresh applications in the new millennium. Philosophical practice is an ancient idea—perhaps the world's second oldest profession—whose time has come again.

PART · II

MANAGING EVERYDAY PROBLEMS

5

Seeking a Relationship

---

*"And yet if every desire were satisfied as soon as it arose how
would men occupy their lives, how would they pass the time?
Imagine this race transported to a Utopia where everything grows
of its own accord and turkeys fly around ready-roasted, where
lovers find one another without any delay and keep one another
without any difficulty: in such a place some men would die of
boredom or hang themselves, some would fight and kill one
another, and thus they would create for themselves more suffering
than nature inflicts on them as it is."*
—ARTHUR SCHOPENHAUER

"To live alone one must be an animal or god."
—FREDERICK NIETZSCHE

Though all kinds of relationships—with family, friends, neighbors,
and colleagues—partially meet the human's hardwired need for social
contact, that need most commonly reveals itself in the search for a love
relationship. Not everyone needs or wants a sustained love relationship,
and some people constantly seek to widen their range of social con-
tacts. But the pair bond is most often the central adult relationship.
People in love relationships need to expend a lot of energy to maintain
those relationships, as we'll discuss in detail in Chapter 6. Many people
not currently in love relationships invest similar amounts of energy to
find one.

If you're among them, you'd do well to find the right relationship to
avoid the devastation of ending relationships (see Chapter 7). Chinese
philosophy teaches that endings are contained within beginnings; a

violent storm brews quickly but cannot last long. Both Christian and Hindu philosophy teaches that we sow what we reap. Taking care from the outset of a potential relationship to make a strong and workable match can go a long way toward assuring a mutually satisfactory result over the long run. Whether the ultimate end of the relationship is death, divorce, or one of the countless other possibilities, the seeds are planted at the outset. Proceeding consciously despite the whirlwind of emotions accompanying any new relationship won't guarantee a smooth course—all relationships have their bumps and bare patches—but it will ultimately give you the best payoff for your investment. This chapter provides philosophical guidance in the hows and whys of seeking relationships.

Finding a life partner in an advanced technological society is harder than finding one in a primitive village, where at least everyone knew everyone else and one would pick a spouse from the relatively few people available. That didn't necessarily yield happy results, of course, but the intimacy of that kind of community provided support for those who didn't make a match or who didn't like the match they'd made. Now our horizons are much less limited, but the trade-off is a loss of community and a fraying of the web of social support that binds groups of people together.

Marshall McLuhan first wrote about the "global village" in the 1960s. Since then the world has become even smaller and more interconnected, with the Internet, cheap and extensive air travel, rootlessness, and globalization bringing the earth's people closer and closer together. McLuhan was talking only about physical space, not cyberspace, and he couldn't anticipate the social effects of globalization. With so many human interactions now mediated by some kind of technological interface, be it a phone or a computer, contact between people loses the intimacy necessary to form individual relationships and in turn communities. The French philosopher Henri Bergson presciently warned about the mechanization of spirit that can accompany technological progress and how it may impede our blossoming as social beings.

"What kind of world would it be if this mechanism should seize the human race entire, and if the peoples, instead of raising themselves to a richer and more harmonious diversity, as persons may do, were to fall into the uniformity of things?"
—HENRI BERGSON

In our accelerated technological society, we become too caught up with worrying about how things work to enjoy a spiritual connection to our world or to other people. Thus the search for someone to share ourselves and our lives with takes on renewed importance.

DOUG

Doug hosted a radio talk show in the wee hours of the morning. He loved his work, including the peculiarities of working the night shift, and found it satisfying. But he was less happy in another important area: he longed for a meaningful love relationship. The awkward hours he kept made it hard for him to meet people and extremely difficult to date anyone. He didn't even come into much contact with people at work (setting aside for the moment the myriad problems of interoffice romance) since the station was run by a skeleton staff during the hours he was there. And when he finished his shift, ready to let loose a little, the rest of the world was eating breakfast and heading off to work.

The irony, of course, was that he interacted with a huge number of people every night: his radio audience. He had a loyal following, and his phone lines were always flooded with callers. He was broadcasting to millions of people who felt they knew him by virtue of bringing him into their homes and cars. Yet he felt as if he knew no one.

Doug serves as an example of the alienating power of technology—and as such he is a kind of icon of modern life. His problem stemmed from a general trend in society toward artificial communities held together by a thin technological thread but devoid of any real social fabric. Technology has unarguably improved human lives a great deal.

One benefit is the expansion of our circle of potential relationships. But the cost of that wider circle is that we get lost in an endless sea of options and possibilities. Without the limits that used to be imposed, we no longer know how to locate or evaluate potential partners.

Taken to an extreme, as we are wont to take things, technology robs us of the kind of genuine community that is crucial for us as innately social beings. You don't have to be as specifically linked to the mass media as Doug was to feel these effects in your life. I bet you can find at least one chat group on the Internet devoted to tossing about complaints about the isolation of modern life—while its members are protracting their isolation through virtual instead of real interactions. But whatever your situation, you are living in a tempestuous world. Relationships provide one kind of safe harbor.

To Doug, all this meant that the technological system that made him a household name exacted an ironic price: loneliness. When he came to see me, he'd already done a lot of work on his problem. Looked at through the steps of the PEACE process, he had already progressed through P, E, and A. He identified his problem: the out-of-phase schedule he kept because of his work, the flimsiness of the social contact he got through callers, and his lack of an intimate relationship. Emotionally, he knew he was unhappy about his isolation and his unfulfilled desire for a meaningful love relationship. Analyzing the situation, Doug saw his available options as unacceptable: leaving his job or staying single indefinitely. He had great job satisfaction and felt no desire to change his employment. He was unwilling to sacrifice important aspects of his career in order to seek a relationship, even though a relationship was also a top priority for him.

Doug was at the point where he needed dialogue to reassess his situation, incorporating all the elements he had examined to find the best philosophical disposition: the contemplative (C) stage of the PEACE process. Through the ages, philosophers have asked people to reexamine their beliefs. That is the essence of the examined life you've heard so much talk about. So I worked with Doug to see if there were other ways for him to conceptualize what he was going through. My job as a philosophical counselor was to get him to reevaluate his own story.

Doug believed that the hours he kept on the job he loved were preventing him from having the relationship he yearned for. He was convinced that those hours stood in the way of his meeting anyone, let alone getting a date. Together we questioned whether this was in fact the whole truth and agreed that there were actually untold millions of people working normal hours who nevertheless struggled with similar frustrations in finding a satisfying relationship. There were surely millions more people working all kinds of odd hours but not lacking for relationships. It was a small but comforting insight: it could well be something other than shift work that was causing Doug heartache. Just knowing that others were in the same situation provided Doug with renewed hope about finding the kind of relationship he wanted, and new energy for attacking the problem. Scientists call this kind of rethinking checking your assumptions, and it's a primary tool for scientific problem-solving. It's no less useful in personal or philosophical problem-solving.

To enter the contemplative way, Doug had to examine alternative explanations of his problem. Did he really need to meet new people to find love—or could he have overlooked someone he already knew? Was he reversing cause and effect—could he have chosen his line of work in part to circumvent relationships? Was there something about him—like shyness—that kept him from meeting people face-to-face (as opposed to facelessly over the airwaves) or from establishing relationships once he did meet someone? To find a way to manage his situation, Doug had to take nothing for granted. Leading that kind of examined life can be initially uncomfortable, since you might not like everything you discover about yourself, but it's better in the long run to know things about yourself, whatever they may be, even if they surprise you temporarily. Only by truly understanding yourself can you recognize your motives, reshape your beliefs, act to achieve your goals, and find more lasting peace of mind.

You might be surprised how many people there are in Doug's position: presentable, intelligent, witty, articulate, established, yet lacking an intimate companion in life. In Doug's case, we discussed a few possibilities that could account for his situation. Could fate be playing a role? Maybe, but Doug didn't like the idea of being completely at the mercy of unknown forces. What about will power then? Doug felt pretty sure that if he decided to order a pizza, he could get one delivered, with his favorite toppings, just by making a phone call. Should finding somebody be that easy? Doug didn't think so. His life partner should be much more special than a slice of pizza. But that doesn't mean she can't be found. Could it be that Doug didn't feel worthy of attracting the kind of woman he really wanted? Or was he in fact doing almost everything right? Maybe his companion was already on her way, and Doug was just fretting over the delay. When you think your pizza's taking too long to arrive, you phone to check up on the delivery. But who can you call when your one true love is taking too long to arrive? And how do you know how long is too long?

I introduced Doug to a couple of relevant insights, one from Taoism, the other from Buddhism. In the first place, Lao Tzu counsels that wanting something badly on the one hand but believing it to be out of your reach on the other will be detrimental to your state of mind. Stand under an apple tree in springtime. You won't see a single apple, and neither will you obtain one by shaking or climbing the tree. Now stand under the same tree in autumn. Ripe apples drop into your hand. There is such a thing as trying too hard, or trying at the wrong time, to obtain your heart's desire. Strive for less desire and better timing. In terms of meeting people, your timing is often best when you're not striving at all. Stop seeking, and you will find. And if you don't find, you won't mind because you're not seeking. This is the art of seeking without seeking. It sounds like a logical paradox, but the Tao isn't bothered by such things—that's how it works.

> *"The five colors will blind a man's sight.*
> *The five sounds will deaden a man's hearing.*
> *The five tastes will spoil a man's palate.*
> *Chasing and hunting will drive a man wild.*
> *Things hard to get will do harm to a man's conduct."*
> —LAO TZU

The second insight is Buddha's: what we experience in life is what we have willed in life—not what we wish or desire or fantasize but what we will. The catch is that what you are experiencing now is a product of your earlier volitions, what you previously willed. You can influence what you will experience in the future by considering what you are willing now, but that process is not instantaneous. It takes time. How much time? Try it, and find out for yourself. With the right practice, you will start living more fully in the present, which means that you will lack almost nothing. You get what you will, not what you want. And you drive away what you want too much.

> *"All phenomena of existence have mind as their precursor, mind as their supreme leader, and of mind are they made."*
> —BUDDHA

In effect, your intimate companion is a manifestation of your mind—as you are of hers. When you can will her to be there, she will be on her way.

As Doug admired the usefulness of these viewpoints, he began to give up lamenting all the places for socializing that were closed when he got off work. Instead he began to ask what was open at that hour. By now he should be a regular at the most popular breakfast place in town, keeping his eyes peeled for another person there who might be thinking, "Wouldn't it be nice to meet people in a civilized fashion, like over breakfast, where you can actually see and hear them, instead of in some dimly lit, deafening basement disco?"

Now that he understood his problem as solvable (rather than as an impossible conflict with his work), Doug was ready to take action, so we talked over some other practical approaches to finding a love relationship, from joining some kind of hobby group to meet someone with common interests to taking out a personal ad specifying someone with unusual hours. Whatever specific steps Doug took, the important thing for him was to avoid philosophical preconceptions that would limit his options.

Philosophical counseling helped Doug through the contemplative stage, at which point he was ready to do something for himself: take steps to meet someone. His disposition had evolved from the problematic ("My schedule is preventing me from meeting someone") to the essential ("I have been using my schedule as an excuse for not meeting someone"). Having effected a fundamental change in his disposition through contemplation, he was now equipped to embark on the adventure he sought. In this essential stage, he no longer had a problem: he was *willing* to follow his heart.

In this fashion, you can shape your own destiny by considering what you will. You may be lost in a maze of wants but not even know it because you don't know what you are thinking about. One purpose of the PEACE process is to uncover whatever program is running in your head and allow you to decide whether you want to change the routine. New Agers tend to take this idea too far, holding that whatever they affirm will be the case. And it would indeed be fun if thinking you were going to win the lottery would make it so. You need to distinguish between what you can change by conscious acts—such as your disposition toward meeting people and therefore your readiness to meet them—and what you can't change by conscious acts—such as the weather.

SUSAN

If Doug had known Susan, he might have considered her experience as proof that meeting lots of people isn't necessarily the road to a satisfying relationship. A successful executive with a big finance company, Susan was the kind of athletic, attractive, well-off thirty-something that models portray in ads for luxury cars. She led a full social life with a wonderful circle of friends, and she never lacked for dates. Yet despite her professional, financial, and social success, she felt she was missing something. She wanted to commit to a long-term relationship and to have children with the right partner, but no one she dated ever seemed to live up to her ideal. Most people she dated didn't even rate a second date with her, let alone a place at the table in her vision of family life.

Susan was bothered by the thought that she ought to have settled down with someone by this point in her life. She had been to countless friends' weddings and commitment ceremonies over the last several years, and now the cycle of christenings and baby-namings was beginning. Her grandmother had mentioned recently that she hoped to live long enough to dance at her oldest grandchild's wedding. Pairing off seemed the normal thing to do. But Susan admitted to being a perfectionist and firmly believed she should be able to settle down without having to settle for anything less than someone who lived up to her standards—all of her standards.

Susan, like Doug, benefited from questioning her own story. Unlike Doug, as she analyzed and integrated the various aspects of her experience, she kept coming back to the same feelings: despite her yearning for a love relationship, she would rather not get involved at all than get involved with the wrong person. As I listened, I agreed with her assessment of her situation, and I didn't see the need to continue questioning it after she had thoroughly reviewed it. It is good to have high standards, and Susan should not measure herself against some kind of statistical average. Among the benefits of philosophical practice are ways to find the essence of yourself and the courage to live it.

Susan prized virtue, in herself and in others. I encouraged Susan to stay with that stance (it is rare enough in today's society) but also to understand what was realistic. All relationships are imperfect. So even if you could find an absolutely wonderful person, there is never a perfect "happily ever after." Also, you can't know in advance who is virtuous, so if Susan wanted to find such a person, she would have to invest time in getting to know people before she passed judgment on them. That was going to take more than one date.

We discussed the possibility of a long-term courtship as a way for Susan to stay true to her search for the right partner without cutting off all candidates before they had a chance to prove themselves. Susan had a kind of reserve that is much less common than it once was. Exploring a relationship slowly could build the basis for something lasting, or at least uncover a sound reason for not continuing the relationship. Susan planned to be honest with potential partners about her need for someone patient in this regard. As a society, we've become so permissive that we lack restraint in most areas, including relationships. Everything moves at a fast and impulsive pace. Whatever the benefits of Internet access or air travel, excessive speed is damaging to courtship. If you are looking for someone to lay a foundation for your house, you want the person who's going to do the strongest, most solid work, not the one who promises to do it overnight.

Susan's thinking about virtue paralleled Aristotle's, who believed that happiness is more than mere pleasure, amusement, or entertainment. He wrote that those things are transient, not enduring, and come from outside the self, while fulfillment comes from within. He called this happiness an "excellence of character" because he saw it as springing from achieving the classical virtues of wisdom, temperance, courage, and justice. (The Christian virtues—faith, hope, and charity—developed centuries later.) For Susan, like Aristotle, fulfillment meant reaching your potential. Aristotle would have added that practicing those virtues means following a middle path. If Susan had been able to confide her troubles to Aristotle, he might have urged her not to compromise her principles but also to make sure that those principles were not extreme.

"If happiness consist of virtuous activity, it must be the activity of the highest virtue, or in other words, of the best part of our nature. . . . We conclude then that happiness reaches as far as the power of thought does, and that the greater a person's power of thought, the greater will be his happiness; not as something accidental but in virtue of his thinking, for that is noble in itself. Hence happiness must be a form of contemplation."
—ARISTOTLE

Susan also responded to the ideas of the Stoics. Despite the popular notion that Stoicism means gritting your teeth in the face of discomfort—taking things "philosophically," as people commonly say—the central concept of Stoicism is to value only that which no one can take from you. Value, then, is found in things like virtue, as opposed to your new fur coat or your Platinum card. For the Stoics, the goal is to maintain power over yourself. If you value something that can be taken away, you put yourself in the power of whoever can take it. Think of how much power a car thief has over those of us who haven't perfected a Stoical attitude. We buy expensive and annoying alarms, struggle to install "the Club" each time we leave the car, and spend a fortune on insurance and private garages—and that's before anything actually happens. If the car is stolen, our emotional distress isn't quantifiable, but it is surely up there on the Richter scale. Cars are plentiful on every street; virtues are rarer in human beings. Susan rightly values her standards and expectations, and the Stoics would give their blessing to resisting forces that influence her to devalue herself.

> *"Nature intended that we should need no great equipment for living happily; each one of us is able to make his own happiness. External things are of slight importance. . . . All that is best for a man lies beyond the power of other men."*
> —SENECA

Susan also had a fatalistic side, so Tolstoy appealed to her. Tolstoy believed in human destiny. Susan, too, felt that fate would play the major hand in any relationship, but she did wonder how much power she had over the situation. Could she find the love she wanted and bring it into her life? Or was it a matter of destiny? As with all philosophical issues, there is no way to answer those questions conclusively. The only way to find out is to live—and even then, of course, you might not find out. Philosophical theories can't be proved like mathematical theorems. Since we don't know the absolute answers, what may matter most is what you believe the answers are, and why you believe as you do.

"The recognition of man's free will as something capable of influencing historical events, that is, as not subject to laws, is the same for history as the recognition of a free force moving the heavenly bodies would be for astronomy."
—LEO TOLSTOY

What it came down to for Susan was waiting for the relationship she had in mind or compromising. If she decided to wait until she found someone who met her lofty expectations, she might find that someone tomorrow—or never. She couldn't dictate the length of the wait. Only she could decide how long she would wait without compromise, and what the nature of that compromise might be.

Both Aristotle and Confucius, who were contemporaries though they lived worlds apart, believed that virtue, like vice, is a habit. Virtue is not an unattainable goal, but rather is well within our grasp. Society does condition us, but at some point we have to assume responsibility for which habits we acquire. Susan prided herself on maintaining virtue and expected a life partner to strive for the same goal. A psychologist might focus on her perfectionism as a fault or a cover for deeper emotions, whereas a philosopher would examine more closely her conception of virtue and how she evaluated it in others. For now, Susan could rest assured that adhering to her elevated standards would offer a path to fulfillment.

We see ourselves reflected in other people, so relationships can help us understand ourselves better. A clearer vision of the self is a goal of philosophical practice, too, which explains why this book has three chapters devoted to relationships. To be fully human, you need to be with others. That's the impetus behind seeking out relationships, as we've seen in this chapter, and the reason we need to work to maintain them once we've found them, as we'll see next.

6

Maintaining a Relationship

"Nothing endures but change."
—Heraclitus

"The whole world is a fire pit. With what state of mind can you avoid being burned?"
—Kao Feng

All structures, from machines and organisms to systems, need maintenance. As a particularly complex, malleable kind of structure, a relationship requires constant repair and preventive measures to keep it functioning smoothly. The balance required to keep a living being functional is continually shifting, necessitating small adjustments all the time. Combine two living beings as elaborate and complicated as humans, and the work to maintain the structure—the relationship—more than doubles. Both individuals have their own needs and wants, and the relationship itself possesses an additional set of requirements.

Both internal and external maintenance may be necessary. Maintaining relationships is hard work, but much is accomplished as a matter of course, without your having to think about it. Your body does a lot of work to keep itself healthy, of which you are usually unaware—regulating temperature and respiration, for example. But at some point you will also need outside intervention, whether it is a flu shot, a course of antibiotics, or a visit to a physical therapist. Keeping a relationship healthy may also require outside assistance, like an anniversary party or counseling. Much of the grease and glue that keep a relationship going come from routine interaction, like deciding together what to eat for dinner, picking up your partner's dry cleaning, or a quick kiss before parting ways in the morning. In between is the middle ground where most of the big issues in any relationship lie and where most of the work must be done.

POWER AND FELICITY

We seek power, as much as we can get. Thomas Hobbes defined your power as your ability to get what you like, or what you think is good for you. Unlike Plato, who saw good as having an ideal Form, Hobbes believed that each individual decides the definition of good. He believed we think something is good because we like it; evil, because we dislike it. To Hobbes, the best we can hope for in life is "felicity," which he saw as a string of happinesses, or the ability to make yourself happy more or less regularly, by getting what you think is good for you. No happiness of his kind lasts very long. (Later on Freud would rediscover that this happiness is transient.) For Hobbes, to maintain one's felicity is the ultimate power.

". . . I put for a general inclination of all mankind, a perpetual and restless desire for power after power, that ceaseth only in death."
—Thomas Hobbes

A well-maintained relationship is a kind of felicity; it is a source of power. Almost everyone agrees that they want relationships to make them happy, or to provide fulfillment. Almost everyone is much less clear about just what they are going to put into the relationship. That dilemma is complicated by what economists call the law of diminishing returns: the more often something happens, the less it is worth. Your first kiss and your early "I love yous" had you seeing stars; now days go by without exchanging either with your partner. Complimenting your partner on a good hair day probably got you lots of points early in the relationship but has become less potent over time.

It is part of human nature to look out for number one. People are self-regarding. Even when we're serving others, we usually do so either because it brings benefits to us or because not doing so would bring detriments. While people sacrifice their lives for others in wartime and other extreme circumstances, this is hardly the norm. Usually, being altruistic meets our own needs too, if not primarily. As an egoistic creature, when push comes to shove, your happiness might come at the expense of your spouse's happiness. That's the setup for a power struggle, and Hobbes would view every relationship as a form of power struggle. The search for the middle ground, for the balance of power, is what I mean by maintenance.

If maintenance is so much work, why bother aiming for that middle ground? The wonderful thing about relationships is that with the proper maintenance, the whole is greater than the sum of its parts. Ideally, both members get support to realize their potential as individuals as well as realizing the potential of the team. If things sour, the tremendous energy drain of an irreparably damaged relationship can also mean that the whole is less than the sum of its parts. (The territory beyond that line in the sand is the subject of the next chapter.) Pooling resources—as in a joint savings account—makes them optimally large. But if one person only deposits and the other person only withdraws, checks are going to start bouncing. Similarly, if only one person in a relationship is performing maintenance and the other is indifferent, their joint account will also wind up with insufficient funds. Overdraft protection might cover everyday necessities, but it won't help when something big comes around.

SARAH AND KEN

For the four years they had been together, Sarah and Ken had found felicity in their relationship. Sarah's friends told her they were jealous of her "perfect" relationship. Sarah and Ken had seen each other through the stressful final year of graduate school and were now enjoying the fruits of their labors, rapidly climbing the ladders at their respective firms. They both worked long hours but loved to hike and ski together when they had free weekends. When Ken was passed over for a promotion, Sarah was able to comfort him and help him prepare for the next opportunity, and when Sarah's mother was diagnosed with cancer, Ken flew home with her every few weeks until the course of chemotherapy was successfully finished.

Sarah and Ken had been living together for three years when Sarah came to see me. She had ended an earlier relationship when the man she was involved with wanted to get married and start a family and she didn't feel ready for that. Now that she was thirty-two and established professionally, she wanted children. But she worried about the difficulty of balancing her career and motherhood and was convinced that for her, a stable marriage was the necessary bedrock for the undertaking.

The problem was that now Ken wasn't ready for that next step. He looked forward to being a father one day but not now, he said (reminding Sarah of Saint Augustine's prayer, "Make me chaste . . . but not yet"). He was very happy with his and Sarah's relationship as it was, and saw marriage as a prelude to parenthood—which in this case it would be. Sarah knew that to become a mother in the way she wanted to be, she would need to change her relationship. But she wanted to change it by bringing it to a new stage, not by taking up with a new person.

Sarah wondered if she should pressure Ken. Could someone consciously get ready for children? Or would pushing him into fatherhood be a big mistake? She thought about getting pregnant and hoping Ken would step up to the plate. But she wanted to start her family more thoughtfully and didn't want to risk the loss of her partner, and single motherhood.

Sarah also acknowledged a financial motive behind her dilemma. As it stood, she pulled her own weight in her relationship with Ken when they measured such things by way of paychecks. But she wanted to be able to stay at home while her children were very young, and she feared what that might mean to both her and Ken. Would uneven incomes create an imbalance in the relationship? Would Ken accept a lower standard of living, even if it was temporary? Would she?

Sarah faced a catch-22 situation. Maintaining the relationship might mean forgoing one of the major things she wanted to maintain a relationship for: having a baby. And having a baby might jeopardize the relationship she wanted to maintain. She chose to evaluate her situation philosophically before making any decisions. She realized that she might not find a solution, but by taking the time to get to know herself and to reexamine the issues, she hoped to find a workable compromise or to simply live through the circumstances until they resolved themselves. Sometimes not doing anything is the right course of action. Maybe all Ken needed was a little time, and then he'd be as eager to be a father as Sarah was to be a mother. She was under some pressure from her parents to provide grandchildren, and she was aware that the longer she waited, the greater the chance that she'd have difficulty conceiving. But she didn't feel any urgency to resolve the issue, and I advised her to take comfort in the process of philosophical inquiry. I suggested she also try to engage Ken in the process, to see if they could find common ground.

Philosophical counseling works by helping to draw out your thoughts on all the important challenges of life and to organize the principles you believe in so that you can act on them. If you share a philosophical framework or can find one to share—be it a religion, a set of rules of your own devising, or some combination of existing systems—it will act as an effective shock absorber throughout a relationship. Most people have dearly held personal philosophies and deep intuitions about how and why things are the way they are, but they haven't formulated this knowing systematically and so can't act in accordance with it. Sarah was working to change that in herself and thought she might persuade Ken to as well.

Bibliotherapy

In Sarah's case, I thought bibliotherapy might benefit her. For some of my clients, I prescribe a book I think might help them. Not everyone likes to work this way, but when the fit is right it can be a big help. For Sarah, I recommended the *I Ching* (I use the Wilhelm-Baynes translation). We discussed this book's central concept: there is a better way and a worse way to conduct yourself in any situation, and if you are wise, you will discover and choose the better way. The *I Ching* is not, as some misconceive it, a fortune-telling device. It is a storehouse of great wisdom, and it mirrors the inner workings of your own mind. This was just what Sarah was seeking, and I thought she would relate to the intuitive approach of the *I Ching*.

The *I Ching* is not meant to be read straight through. Rather, you use it by asking it a question that you want to answer. You're really asking yourself; the *I Ching* helps you find your answer within. A simple system of tossing coins (described in detail in Appendix E) points you to a particular section of the text. The text consists of sixty-four sections, each of which corresponds to one of sixty-four possible hexagrams, obtained by tossing the coins. Sarah obtained hexagram 37, "The Family," where she read, among other things:

The foundation of the family is the relationship between husband and wife. The tie that holds the family together lies in the loyalty and perseverance of the wife. . . .

Another passage advised:

The wife should not follow her whims. She must attend within to the food. Perseverance brings good fortune.

The text went on to explain:

. . . in this way the wife becomes the center of the social and religious life of the family, and her perseverance in this position brings good fortune to the whole house. In relation to general conditions, the counsel given here is to seek nothing by means of force, but quietly to confine oneself to the duties at hand.

Together, Sarah and I discussed possible interpretations of this hexagram. Sarah found the passages quoted here especially meaningful, and ultimately she concluded she should not force parenthood on her husband. She supposed that if she persevered as a good and loving wife, Ken would realize that she would also make a good and loving mother. Then their parenthood would be natural, not coerced.

Thus the *I Ching* helped Sarah help herself through contemplation, giving her a philosophical disposition that would get her what she wanted but through an uncontrived course of action that would bring no disasters in its train.

The loveliness of the prose and the practical nature of its wisdom is enough of a reason to check out the *I Ching*, but another reason I suggested it to Sarah is reflected in the other common name for this ancient Chinese text: *The Book of Changes*. As Heraclitus observed, and Sarah was experiencing, "Nothing endures but change." The nature of a relationship is maintaining something in the face of change. That's not to say that a relationship doesn't change, because it must. A marriage isn't the same at the golden anniversary as it was fifty years earlier. There have been various phases, cycles, crises, and resolutions along the way. Malleability is the key to the survival of the relationship.

A roller coaster is one thing, but riding one brings changes every second. Going up and going down are two very different sensations, even though you're still on the same roller coaster. Sarah knew that the car she was riding in with Ken was headed for the top of the hill, but she wasn't sure what was on the other side of the slope: a family, a relationship without children, or the end of the relationship. In any case, change was coming, and she wanted to find the best way through it.

Chinese philosophy teaches that you may not be wholly responsible for a situation but that you do have an obligation to find the best way through once you are in it. According to the *I Ching*, if you take the worse way to accomplish what you want to accomplish, there will be problems further down the road (as Sarah feared when she contemplated an "accidental" pregnancy). Always seeking the best way will allow you not only to make the best of a good situation but also to make the best of a bad situation.

Machiavelli isn't generally a philosopher you'd trust for guidance on love relationships. But in *The Prince*, he did lay out a helpful theory about change. He thought that everything that happens to us is due to either destiny or ability—with just about a 50–50 split between their influences.

"I am disposed to hold that fortune is the arbiter of half our actions, but that it lets us control roughly the other half."
—NICCOLO MACHIAVELLI

In any situation, you must take credit for what you can do, given your own resourcefulness. But you must also acknowledge the forces of the universe over which you have no control. Sarah need not be lulled into inaction by a fatalistic outlook. But she must also bow to things beyond her ability to influence. Just how Machiavelli's *fortuna* and *virtù* (destiny and ability) would combine when it came to changing Ken's attitude toward becoming a father was what Sarah had to pinpoint. Chinese philosophy holds that you need to ascertain what kind of situation you are in and choose a course accordingly. In a storm, you batten down the hatches until it blows over. With a favorable wind, you unfurl every sail. Sarah felt that whatever the ultimate destination, the trip would go better with a pair of navigators (reading from maps laid out by world-class cartographers) than with a lone captain.

TONYA

My colleague Christopher McCullough had a client who came to him worried that she had married the wrong man. But then, Tonya hastened to add, she always started to feel dissatisfied two or three years into a relationship. Her marriage marked the first time she felt committed to a relationship over the long term, but still she was getting those old feelings. Had she made a terrible mistake? Despite her doubts, she said she loved her husband dearly. She wanted to remain committed to her marriage yet felt trapped in it. We'll review this case in terms of the PEACE process to get a picture of how the philosophical counselor helped Tonya.

Her problem was clear: fear that her relationship was failing. Growing out of that central concern, Tonya had additional fears: that she was incapable of having a lasting relationship and that a lasting relationship meant feeling trapped. She was equally clear about her associated emotions—in addition to fear, she felt guilt, sadness, insecurity, anxiety, and disconnection.

It was in her analysis that Tonya ran into trouble. She expertly described her psychological history, picking out various themes so well that it was clear she'd spent time in therapy. (Again, psychological counseling can be helpful in this way: helping you to know yourself.) But when it came to her options for action, she saw few possibilities: stay married whether she liked it or not (the path her parents took) or concede that she wasn't meant for a long-term relationship, break up the marriage, and give up the idea of ever settling down permanently. The option she was overlooking, despite its obviousness, was staying happily married.

Her counselor guided Tonya through the contemplation stage. When he asked her to define what she meant by being committed to her relationship, she said it meant she had taken on an obligation and was living up to her family's and society's expectations. This reflected a philosophical disposition—marriage as obligation to others—that clearly was not serving her well.

Presented with an alternative way of looking at marriage, Tonya agreed that marriage could also be a commitment undertaken because she wanted to be in a serious, exclusive relationship as well as to meet someone else's wants and needs. She realized that remaining with or being unfaithful to her husband was a decision she would make, not a condition forced on her. She said she felt the only reason she didn't cheat—since she had been tempted—was that she didn't want to hurt her husband. In her mind, that had felt as though he were making her do something she didn't really want to do. But now she realized that in rejecting the possibility of being with other men, she herself was making a choice. Acting on her own, she held as a top priority not harming the man she loved. She really didn't want to hurt him.

Tonya concluded that commitment was not a loss of freedom but rather an exercise of freedom. Tonya restructured her philosophy of relationships around this key insight, and that shift made all the difference to her. Understanding that the option of leaving was always there—and that she was continually making the decision to stay—calmed Tonya's fears. She no longer felt trapped, because she knew there was an exit but that she chose not to use it.

"No man is free who cannot command himself."
—PYTHAGORAS

Tonya and her husband still had in front of them the work all couples must do to maintain a happy relationship. But without acknowledging a fundamental commitment, there is nothing to build on. This first step was the crucial one for Tonya.

In Tonya's case, what she learned about her relationship problem spilled over directly into other areas of her life as she reached equilibrium. Most significantly, she committed herself wholeheartedly to her business, which had been foundering for years. Freed of her fears of being trapped in success along with her fears of being trapped in a relationship, she now poured herself into her business, which suddenly took off. Tonya's a great example of how the difficult and sometimes painful work you do in fashioning a philosophy to carry you through one circumstance will often stand you in good stead in other areas of your life.

EXTERNAL AUTHORITY

Hobbes wrote that if people recognize a common strong authority, they will be obedient to that authority and will get along with each other. Boundaries everyone respects will limit conflict. Then again, in the absence of such an authority, and without agreed-upon limits, conflict will always escalate. Hobbes was writing about civil war, but the ideas apply on more personal battlegrounds as well. The external authority limiting conflicts in relationships can be ecclesiastical, civil, or philosophical.

In days gone by, there was a stronger secular authority keeping people in marriages: the social stigma of divorce. Not all marriages were happy, of course, but even in difficult relationships there was a strong incentive to get along because the alternatives were unthinkable. Some religions, like Catholicism, still carry that kind of authority with their followers, though you can see in the public battles over the annulment process and the number of lapsed Catholics that Catholicism's rules on the subject are increasingly difficult to abide by.

We now have independence from that kind of external authority, for the most part. The resulting individualism is a boon in some ways, but it also initiates a certain amount of social anarchy. Without a rule book to play by, people don't know how to be with each other anymore. Philosophical ideas, whether promoted by academic study, explored in counseling, or formulated on your own, can act as an outside authority and, when those ideas are mutually respected, can keep the peace.

> *"During the time men live without a common power to keep them all in awe, they are in that condition which is called war."*
> —Thomas Hobbes

Without common constraints on relationships, our focus is no longer on maintenance but on whether the relationship works at all. If the relationship is seen in a negative light, the next step is often to end it rather than work to improve it. Just as more and more people are leasing cars rather than making the long-term commitment of buying, they seem in some sense to be leasing relationships. A lease is well and good when it means a shiny new car in the driveway every thirty-six months, but it will never produce the kind of constant care and continual upkeep that a lasting relationship requires. Why bother with an oil change if there's only another year on the lease? Why go out of your way to please a demanding spouse if you can get divorced instead?

This "for better, but not worse" approach to relationships will never yield intimacy or deep connection. The other problem is that without strong commitment, there is less reason to be civil in the midst of conflict. That creates the potential for real ugliness: you know where your partner's buttons are, and you're not afraid to push them. When all's fair in love (as in war), someone may get seriously wounded.

So here's to détente. The trade-off for relative peace may be a lack of complete liberty from authority, but the return makes the investment worthwhile. Hobbes would agree that the wise course is to give up some of your power in exchange for getting some security out of cooperating with others. As social animals, we are stronger in groups and when we are working together toward common goals. Everything you needed to know, you learned in kindergarten: take turns, and share.

SEAN

My colleague Richard Dance counseled Sean, who had been dating Patricia for nearly eight years. They had broken up and gotten back together more than once over that time and had recently become engaged. Sean felt he had finally found "the right one" in Patricia but also felt he had some issues to work out that would be better settled before getting married, so he sought philosophical counseling.

His biggest complaint about Patricia—whom he described as attractive, witty, honest, and self-reliant—was that she didn't take his advice. For example, she had recently graduated from law school and was putting all her effort into landing one particular position. He felt it was a mistake to keep such a narrow focus and wanted her to branch out in search of more job opportunities. When it came to maintaining their relationship, he felt she didn't accept his criticism there either.

As Sean and his counselor outlined his concerns, it became clear that his underlying themes were having strong opinions, seeing the world in black-and-white, and seeking control. He also tended to overanalyze things yet not quite trust his own decisions (taking, as just one example, the eight-year trial period he found necessary before deciding that Patricia was the woman for him).

In philosophical counseling, Sean learned a meditation technique from the Hindu tradition where the objective is to mentally replay a recent event in which any strong feelings came up, but to do so objectively, without emotion, analysis, or judgment. During the next few weeks, Sean practiced reviewing—but not reliving—conflicts with his fiancée, putting himself in the position of a disinterested third party regarding his own and Patricia's behavior. Soon he noticed that when a new situation came up in which he would normally have launched into delivering his opinion, he caught himself before he spoke. He could then calmly decide not to criticize her and realize that she would make her own decisions.

As Sean's new insights changed the way he interacted with Patricia, his counselor added a second assignment: to evaluate whether his experiences bore out the wisdom of Lao Tzu and Heraclitus about the coincidence of opposites. Heraclitus wrote, "Disease makes health pleasant and good," which may as well have come from Lao Tzu's pen: "Difficult and easy define each other." Both ancient Greek and ancient Chinese thinkers held as a central concept that opposites are interconnected, relying on one another to complement their mutual existence. That shed a whole new light on holding conflicting opinions. As he identified examples of this phenomenon in his own life, Sean came to see that some of his opinions had been important to him only because they were different from some other opinion—often Patricia's! As Sean observed, color contrast is one of the things that makes art visible. But clashing opinions don't make relationships workable.

Reframing his differences with his fiancée this way and taking a step back from snap judgments (and the desire to impose them on others) made an immediate difference in Sean's relationship. He and Patricia didn't argue as much in the first place, and when they did it tended to be a substantive discussion that ultimately strengthened the foundations of the relationship, not a flash fire threatening to burn out of control. They were having a great time planning their wedding together, and they could both enjoy certainty that the commitment of marriage was the best thing for both of them.

Sean did the wise thing in seeking to resolve outstanding issues before making a final commitment to a relationship. The dispositions he formed in counseling will serve him well, but the most important disposition—the willingness to invest energy in making a relationship work if you expect it to last—he brought with him.

> *"They would not know the name of right*
> *if its opposite did not exist."*
> —HERACLITUS

NORA AND TOM

Nora and Tom had the kind of relationship that would have been more successful in another era, when both of them would most likely have shared a vision of how the relationship should be. But after many years in a traditional setup—husband as breadwinner, wife as homemaker—Nora and Tom now held conflicting views of what they could expect from their marriage.

Nora had dropped out of college when she gave birth to their son, Nicky. Now that Nicky was in first grade, she wanted to complete her degree and work full-time. She had enjoyed being at home to raise Nicky, but she didn't like being completely dependent on her husband financially. Tom opposed the idea of Nora working outside the home; in fact, he wanted Nora to have more children and stay home with them. He made good money as an accountant, so they didn't need another income to get by.

Nora felt dominated by Tom and told me that he seemed to feel threatened by her wanting to graduate and find full-time work. Furthermore, no matter how much work she did around the house, Tom was never satisfied with the job she'd done. And although they could afford it, he wouldn't allow Nora to hire a housekeeper, insisting that she clean the house herself. Nora's analysis was that he wanted her to clean the house more than he wanted the house cleaned. Tom was not a very involved father despite his interest in having more children. When Nora asked Tom to supervise Nicky's homework so that she could spend some time studying in preparation for going back to school, he let Nicky watch television instead.

Despite the lack of recognition she got from Tom, Nora wanted to maintain the relationship. They did have some wonderful times together as a couple and as a family, including enjoyable vacations. And to Nora, the alternative to preserving the marriage—single parenthood and living with her mother—was extremely unappealing. The price of keeping the relationship seemed high, but Nora felt she was choosing the lesser of two evils in "marital servitude." Her goal was to stay together long enough to improve her financial situation so that she could stay—or leave—on her own terms.

"We must as second best . . . take the least of the evils."
—ARISTOTLE

Power Struggle

Nora and Tom were engaged in a classic power struggle. Hobbes's advice would be to work to find a balance of power, but Tom was currently unwilling to make any changes. Since Nora's choice was to live with Tom as the dominant partner, her task became to adopt the best philosophical disposition toward her situation: the most beneficial way of looking at and thinking about it. On a very basic level this might mean considering whether she should continue to keep up with the housework if her efforts in that arena didn't earn her Tom's appreciation or his support for other endeavors. For the time it took her to get through school and find a decent job, Nora would be facing opposition from her husband. So she should find the path through the minefield that made her feel least stressed, endangered, or unhappy.

To that end, I counseled Nora to explore her part in the relationship. To make her course as smooth as possible, she needed to pinpoint what she did that provoked Tom, whether or not Tom's response was justified. Critical thinking shows that you can be causally responsible for something without being morally responsible. You may be provoking someone without knowing and without intending to do so. Figuring out the causes of her confrontations with Tom could help Nora keep them to a minimum. Even though she was not necessarily at fault, more awareness of cause and effect might minimize the cycle. Once Nora understood the effects of what she was doing, she could decide whether or not she was willing to change it. As Leibniz wrote, everything happens for a reason. If you want to exercise some control over what is happening, you have to understand the reasons behind it.

> *"There can be found no fact that is true or existent . . . without there being a sufficient reason for its being so and not otherwise. . . ."*
> —Gottfried Leibniz

Master and Slave

Nora also drew insight from Hegel's theory of the master-slave relationship. To all outward appearances, it is the master who oppresses the slave. But Hegel saw that the master was also dependent on the slave, not just economically but also emotionally. This dynamic actually gives the slave a great deal of power in the relationship. Hegel sees power as existing in two aspects: for someone to be powerful, someone else has to be powerless. Given that Hegel saw every relationship as some variation on master and slave, he's not an altogether appealing philosopher when it comes to human interaction. But his insight into what happens when there is an imbalance of power is instructive.

Masters derive their sense of self-worth from the oppression of someone else; the only way they can keep their own fear of being enslaved at bay is to enslave another person. This was Hegel's explanation for why masters are so reluctant to liberate their slaves. In the case at hand, Tom's security hinged on Nora's insecurity; he was dependent on her dependency. Nora might be able to draw strength from simply recognizing her own power in the relationship. She had to exercise that power wisely, however, because it corrupts. If she reversed the situation and became the master (for example, by threatening to divorce Tom if he didn't grant her more liberty), she could become the one holding the master's precarious power.

If Tom were willing to undertake philosophical counseling, he could also learn from Hegel. Anyone preoccupied with fears of being toppled from power is in an unstable position. With power-sharing, you can draw strength from a relationship rather than expending all your energy maintaining dominance. Physicists define power as the amount of energy expended in a given time. Energy comes in many forms: nuclear, chemical, biological, emotional, and intellectual, among others. If well maintained, a relationship is a power generator too—a source of clean, safe, and abundant energy.

JUST YOU AND I

The Jewish philosopher and theologian Martin Buber divides relationships into two types: I-Thou and I-It. The first represents a mutual give-and-take between equals, while the second is proprietary and manipulative, as between a person and an object. A healthy relationship needs mostly I-Thou interactions, but we often make the mistake of treating other people like things, creating an I-It dynamic. That's another way of setting up an imbalance of power that will lead to strife. Kant covered similar territory in calling on us to treat other people as ends in themselves, not means to our ends. Rather than controlling others for our purposes, we should appreciate them as people with their own sets of goals. To know a thing by its opposite, consider the ruthlessness of Machiavelli's *The Prince*, which Bertrand Russell called a "handbook for gangsters." Machiavelli's techniques illustrate the I-It relationship perfectly when he tells how to govern people by making them fear you, how to enrich yourself at others' expense, and how to cling to whatever power you obtain. If Dictator is the job title you're after, Machiavelli straight up may well work for you. But if your interpersonal interests lie more in the happily-ever-after realm, you'll do better with Buber and Kant. They would both agree with Hobbes that a balance of power is the key to a successful, peaceful relationship.

CIVIL WAR

Hobbes held that if we knew in advance the worst that war could do to us, that knowledge would be an effective deterrent. He was writing specifically about civil war, as opposed to international war, because he felt that closeness gives a greater potential to wound. Consider the constructive relations the United States developed with Germany and Japan, beginning immediately after the terrible events of World War II. Then compare that with the overheated emotional reaction you are still bound to get by bringing up the Civil War just about anywhere in the American South. The War between the States is still being fought at home on many levels, almost a century and a half later, while our foreign enemies of fifty years ago are now our friends. Similarly, family feuds are more bitter and protracted than neighborly disputes.

You may be your own worst enemy, but next in line is the person you share an intimate relationship with. Living with someone gives you a lot of information about how to make that person unhappy. Your partner knows right where your soft underbelly is exposed. If people knew the emotional, legal, and financial devastation of divorce ahead of time, they might vindicate Hobbes's theory of deterrence by staying together. The painful upheaval that accompanies the end of most love relationships is perhaps the best reason for working hard to maintain a healthy, happy relationship. Sometimes those efforts fail or are futile no matter how sincerely and consistently they are undertaken, so we'll discuss ending relationships in the next chapter.

Socrates might have been taking into account the ability of those closest to us to hurt us the most, as well as love us the best, when he formulated his symmetric ethic: you have a capacity to do a certain amount of good, which is always accompanied by the ability to do a similar amount of evil. The more passion people have, the more they can attract or repel each other. Think of Elizabeth Taylor and Richard Burton. Married twice and divorced three times from each other, they alternated between extremes of attraction and repulsion. When traveling together, they had to rent additional hotel rooms surrounding the one they were staying in and keep them vacant because their quarrels were audible through the walls. They couldn't quite extricate themselves from the relationship even when they tried. They may have been "made for each other," but their attraction came twinned with repulsion. The opposite of passionate love is not hatred; it is indifference. Even if you are really in love, and even if the feeling is mutual, you still aren't guaranteed happiness. If you want to avoid painful endings, you'll have to use your power in a relationship wisely and keep up the necessary maintenance.

> *"I only wish it were so, Crito, that the many could do the greatest evil; for then they would also be able to do the greatest good—and what a fine thing this would be!"*
> —SOCRATES

7

Ending a Relationship

"Be wisely selfish."
—DALAI LAMA

*"Take care with the end as you do with the beginning,
and you will have no failure."*
—LAO TZU

No relationship is perfect, because people are imperfect. Diligent maintenance is the key to any lasting relationship, as we saw in the previous chapter. The ravages of the end of a marriage or any established relationship should provide incentive for giving your all to making it work before you give up on it. In many relationships, however, maintenance fails or is too one-sided, and the ties that bind, break. Defining the limits of maintenance is one of the tasks of being in a relationship, but once you are at or over that borderline, philosophers have a lot to say about how you might proceed.

JANET

When Janet came to me, she was in crisis—personally, emotionally, and philosophically. She and her husband, Bob, were "spiraling unhappily toward divorce," she said, and she was trying to decide if she should go home that evening or spend the weekend at a hotel deliberating over whether she should finally leave him permanently. They were both successful professionals, and by mutual agreement they had no children. They lived in a beautiful house near the beach and with their combined salaries enjoyed a very comfortable lifestyle.

Bob, Janet told me, was extremely hard to please. But she was intent on giving him everything he required from her. She took responsibility for taking care of their home, on top of her long work hours. She fastidiously maintained her appearance, working out almost daily and carefully choosing stylish clothes that flattered her slim figure. She compromised on her preferences whenever they had a conflict, no matter how large or small. When she had her heart set on a vacation touring Italy, he chose two weeks on the beach in Barbados instead—and she made all the arrangements for the trip. When she had an opportunity for a promotion that would have required relocating to a new city, she passed it up because Bob was reluctant to move. If she was in the mood for Japanese food for dinner, and he had a yen for Italian—they went for Italian.

But the more Janet gave, the more critical of her Bob became. The more she did for him, the less appreciated she felt, and she felt the marriage falling apart. He complained that there was nothing to do in Barbados; he criticized her lack of ambition in her career; he didn't like her new haircut; and he thought the maid she hired didn't fold his shirts correctly. It seemed that nothing Janet did was right.

Yet Janet remained unsure about her next step. Should she leave Bob? As she neared the end of her story, Janet told me that her first marriage had failed, and she worried that she hadn't worked hard enough to save that relationship. Now she found herself in another unhappy relationship, fearing the stigma of failing in marriage again. She didn't want to be a two-time loser and vowed not to make the same mistake twice.

Janet had been through therapy around the time of her first divorce, as well as when things had begun to fall apart with Bob. And though she now understood the role her relationship with her father had played in her relationship with her first husband, she still didn't know what to do about Bob. The most recent psychiatrist she had talked to wrote her a prescription for an antidepressant, but since it didn't seem to make a difference in her mood, she didn't renew it when the original supply ran out.

She and Bob had even been in marital counseling together, briefly. The counselor had suggested some new communication and negotiation techniques, but Bob had scoffed at the idea of them as soon as they stepped out of the office. Janet didn't think they had any problem making themselves clear to each other, but she was willing to give just about anything a try. Bob had been unwilling to go back after the first few sessions, however, saying it was just a waste of time and money.

Janet felt that traditional psychological approaches hadn't helped her in the wake of her divorce. Nor was she finding any help there in her current situation. So when she heard me on the radio discussing philosophical counseling, she called me immediately on her car phone to make an appointment. We had our first conversation that same afternoon—the day she was deciding whether or not to leave Bob. She knew she was in a crisis situation but couldn't see her way through.

Before I agree to work with people, I assess whether they are good candidates for therapy with me. Janet's was a case where philosophical counseling was quite appropriate. She was upset yet still in control. She was functioning well at her job, sleeping well at night, and accomplishing all the many things she had to do in her life at her customary level. People experiencing severe distress or dysfunction in the ordinary routine of their lives may need to see a physician or psychiatrist and get some temporary help from medication before seeing a philosophical counselor.

Selfishness as a Virtue

The first things I discussed with Janet were Ayn Rand's ideas on the virtues of self-interest. Though Janet had never read *The Fountainhead* or *Atlas Shrugged*, she agreed that she was now being selfless to the point of her own detriment. And although she prided herself on her altruistic nature, she responded to Rand's idea that you can't really help others unless you are secure yourself. If Janet decided she had given as much as she could to this marriage but that Bob hadn't responded and wasn't likely to respond, Rand would say she had an obligation to preserve and protect herself by leaving the relationship. When an emotional investment doesn't pay off—and in fact requires ever-increasing contributions—Rand would advocate cutting your losses while there's still something left to reinvest elsewhere.

"I swear—by my life and my love of it—that I will never live for the sake of another man, nor ask another man to live for mine."
—AYN RAND

Rand is commonly associated with libertarianism and is known for valuing rationality and intellect. But she is not alone in asserting the value and morality of self-interest. As the Dalai Lama says, "Be wisely selfish." The Mahayana tradition of Buddhism holds that everyone should attain enlightenment and that when you get there you should turn back to help others reach the same goal. But you must be in a position of clarity yourself before you can help others become clearer. When selfishness arises from enlightened self-interest, it is a constructive force. When it springs from vanity, egotism, or narcissism, it is destructive.

These thoughts, too, hit home with Janet. As a businesswoman, she well knew the value of logic and straightforward reasoning—but also the power of intuition. She realized that she could respond to her marriage problems both intellectually and intuitively, possibly reaching the same conclusion with either method. While I do not impose my views on my clients, it is my job to be an advocate for their self-interest. For Janet, that was a component missing from all her earlier experiences with therapy. She didn't want to be told what to do, but she did want guidance in taking specific actions—not a blank slate, or a wall of "That's interesting . . . continue" responses.

As her advocate, I recommended further philosophical work to resolve her issues with her father and her first marriage. I felt she would risk following the same pattern in a future relationship if she couldn't get rid of the psychological baggage she was carrying. In living an examined life (or working with a philosophical counselor), you should be looking at what is missing as well as what is there. For Janet this meant realizing that she was lacking interpersonal resolution in some areas of her life and needed to settle past unfruitful relationships in her own mind. Then she would become liberated to discover who she was, and that authentic person could assert herself in a future fruitful relationship. She needed to enact a successful present in her quest for fulfillment instead of reenacting an unsuccessful past.

As with many of my clients, Janet drew comfort from knowing that she was not alone in her line of thought. She had grown uncomfortable putting so much into her marriage as she got less and less out of it other than criticism and unhappiness. Though she wasn't yet sure she had reached the limit, she agreed that at some point there is a line in the sand separating constructive selfishness and destructive selflessness—and loved the fact that some of the great thinkers of our time had reached similar conclusions.

I only counseled Janet for that one session. While that is not unheard-of, philosophical counselors more commonly see people at least several times. Most of my clients see me for short-term therapy—usually not longer than three to six months. Some people, like Janet, need help through a specific crisis and also need to address longer-standing problems.

The PEACE Process

Here's how Janet's case shapes up in terms of the PEACE process.

First, the problem: Faced with a failing marriage, Janet needed to decide whether to go home that night and try to make things right with her overdemanding husband or to check into a hotel for the weekend and spend the time reflecting on her own.

Second, emotions: Janet felt frustration, despair, and anger at the prospect of going home. She knew she couldn't really please Bob even if she sincerely tried, and found it hard to reconcile her professional success with her marital failure. But Janet also felt fear and hopelessness at the prospect of checking into a hotel. This might signal the beginning of the end of her second marriage; her first marriage had already failed for similar reasons. She did not want to bear the stigma of being unable to maintain a marriage.

Third, analysis: In her conversation with me, Janet explained that she had been undervalued by both her parents and in particular had never received much approval or recognition from her father. In consequence, she felt undeserving of her father's love and assumed as a young child that his lack of love was due to some glaring deficiency on her part. My colleague Pierre Grimes calls this kind of false belief about oneself a "pathologos": it colonizes one's ability to succeed and instead makes one's failure self-fulfilling. Janet's pathologos, "I do not deserve my father's love," became translated in subsequent marriages into "I do not deserve my husband's love." Her pathologos made sure she would marry exactly the wrong kind of man and, worse, would blame herself when the marriage failed. She was caught in a trap of her own devising. Both horns of Janet's dilemma—either return home or check into a hotel—served potentially to reinforce the pathologos: either way, she would not receive her husband's love and could therefore declare herself undeserving of it.

The role of Socrates, as portrayed by Plato in the dialogue *Thaetetus*, is that of a philosophical midwife. We are all pregnant with ideas and need the midwife to help bring them into the light of day. But the philosophical midwife also helps us distinguish between those ideas we have conceived and those—like the pathologos—disguised as our ideas that in fact are harmful impostors.

"But the greatest thing about my art is this, that it can test in every way whether the mind . . . is bringing forth a mere image, an imposture, or a real and genuine offspring."
—PLATO

Fourth, contemplation: Recognizing that you hold a false and destructive belief about yourself is one thing; replacing it with a true and constructive belief is another. In general, you do not reverse deeply ingrained convictions merely by reconceptualizing them. A pathologos is reinforced by experience. The only way to reverse it is to accumulate a different quality of experience—guided by constructive beliefs about yourself—and replace the edifice of self-destruction, brick by brick, with one of self-affirmation. This is accomplished literally one day—even one hour or one minute—at a time. Janet's pathologos needed to be replaced with a belief such as "I was deserving of my father's love, but he was incapable of loving me because of his own problems," which would then lead her to believe, "I am deserving of a husband's love, so I must find a husband who can love me."

By exercising this new idea, Janet would be able to attract a husband who could and did love her. But the first step is always the most difficult and requires courage. The pathologos is disguised as an old friend, and leaving it behind might seem like abandonment. In fact, it is one's worst enemy, and must be abandoned if one is to lead a fulfilled life.

Fifth, equilibrium: Now Janet understood not only that her impulse to check into a hotel was self-preservative, but also that she had a self worthy of preserving. By spending some time alone, having no one around to boost her ego but no one around to demean it either, she could enjoy the equilibrium of delicious solitude necessary to recognize her worthiness—and ultimately to attract others who would recognize her worthiness too.

At the end of our session, Janet told me that I had given her a lot to think about and that she now had confidence in her ability to make the right decision. If nothing else, I knew it would now be possible for her to stop "spiraling unhappily toward divorce." She might even begin spiraling happily toward it! While a lasting marriage is usually best for all concerned, it may sometimes be preferable to get divorced for the right reasons rather than stay married for the wrong ones. If you begin to discover yourself philosophically, your life may change as a result. Sometimes that change can be unsettling, and you need courage and determination to see it through. But such philosophical growth also leads to philosophical self-sufficiency, and that is what allows you to be true to yourself. This is what Janet wanted, and I think she will get there.

Janet's story provides an interesting contrast to Nora's, the woman who wanted to finish her degree over her husband's objections, in Chapter 6. Both women were struggling with critical, demanding men who were unsatisfied no matter what efforts their wives made. But the different outcomes illustrate how different philosophical solutions can apply to similar problems, depending on the people involved. We all have our own unique philosophical outlook, and no two people respond in the same way for the same reasons, even given similar circumstances.

LARRY

Larry was also struggling with the possible end of a relationship. Married to Carol for nearly twenty-five years, he had two young adult children. He had been faithful to his wife for all those years, and they prided themselves on the partnership they had formed when it came to raising and providing for their children. They both had flourishing careers, though Carol worked from home and for many years had worked part-time in order to spend more time with their children. Larry respected his wife, but now that they had an empty nest, they found they just didn't have much in common any more.

Tellingly, when Larry approached Carol to talk seriously about their commitment to each other, she told him she didn't want to listen to him ruminate on the subject anymore and suggested that he pay someone else to listen to him. Part of the reason for having a love relationship is to participate in a continuous dialogue, so Carol's response demonstrated that this key element of their relationship had broken down. Home is not only where the heart is, and where they have to take you in, but also where people are interested in what you have to say—interested in you as a human being without ulterior motives, valuing you because of who you are.

Larry had never been to a psychiatrist or psychologist and would have resented even the suggestion of extended therapy. He came to me, then, at his wife's suggestion, simply looking for someone to talk to as he considered whether he should leave her. He definitely did not want to discuss his feelings—much less his childhood or his patterns of behavior. Like most of my clients, he was looking for someone who could help him articulate his worldview (that is, his personal philosophy) and examine his choices to make sure the actions he took were consistent with his beliefs and values. That task is not always as simple as it sounds.

Larry and Carol were both principled, loyal people and perceived themselves as operating within a serious ethical framework. They were not religious people, but they had formulated their own moral precepts and abided by them. Now as Larry considered choosing an end (divorce) not necessarily in keeping with his principles (marriage as a lifelong commitment), he was asking himself if there was ever a time to change the rules he held as absolutes. When blindly obeying a rule starts to inflict damage, it may be time to change the rule.

Marriage vows are usually taken "till death do us part"—that is, for life. But suppose you discovered sometime after the honeymoon that you had married a psychopath or a sadist who had cleverly deceived you and who now might really harm you or ruin your life. In that dangerous case, maintaining your wedding vows would probably do you more damage than breaking them. Now consider a more trivial case, where you quarrel with a sibling or a close friend and then vow, "I'll never speak to you again as long as I live!" After a short while, you really miss that person, who also misses you. Keeping your vow never to speak to him or her again would probably do you more damage than breaking it, so you make the phone call.

Larry's case falls somewhere between these extremes. Two people may share a wonderful marriage for many years, all the while still growing as persons and all the while intending to keep their vows. Yet the day may arrive when they have both outgrown it, in which case maintaining the marriage would do more damage than dissolving it. If only one spouse feels this way, they may both be in for a difficult time. But if both feel this way, which is less common, they can actually preserve their love by letting go of their marriage. This, I believe, is what Larry and his wife managed to accomplish.

Duty

Kant thought that moral duty must be performed for its own sake and that morality comes from reason. Like Kant, Larry was a moralist, so Kant's approach fit right in with his. Kant wrote of several "perfect duties" humans have, and his list of things never to do (e.g., lie, kill) sounds more or less like the Ten Commandments. He also notes the "imperfect duties" we have, one of which is to improve ourselves. Unlike perfect duties, which are universal, imperfect duties are situational. Applied to Larry's case, this might mean that although marriage (a mutual obligation) is a serious commitment that ought not to be violated, if that mutual sense of obligation has ceased, then maybe staying within the marriage would not benefit either Larry or his wife—thereby violating the "imperfect duty" they both have to improve themselves.

> "To secure one's own happiness is a duty, at least indirectly; for discontent with one's own condition, under a pressure of many anxieties and amidst unsatisfied wants, might easily become a great temptation to transgression of duty."
> —IMMANUEL KANT

William Ross's theory of prima facie duties would lead Larry to similar conclusions. Ross wrote that we all have a list of commitments that "at first glance" (prima facie) are all equally binding but that in practice these commitments are sometimes going to conflict. He holds that different situations call for different priorities and that every case must be decided on its merits. So while his children were young, Larry's primary commitment may have been to them, and he might have maintained the marriage to support them emotionally. But now that the situation had changed (his children were grown), his first obligation might be to support his own emotional growth by leaving the marriage.

> "When I am in a situation, as perhaps I always am, in which more than one of these prima facie duties is incumbent on me, what I have to do is to study the situation as fully as I can until I form the considered opinion (it is never more) that in the circumstances one of them is more incumbent than any other. . . ."
> —WILLIAM ROSS

Change

If Larry had been a more intuitive sort, I might have discussed the *Tao Te Ching* with him. Like the *I Ching*, this ancient Chinese text takes as its premise that everything changes and that to understand change, you must understand the nature of the laws—the Way—driving that change. Another of its underlying principles is that there is always a choice between a better and a worse way to do things. Ideally, the best choice will leave you blameless. Blame is a crucial concept in Chinese philosophy, playing a similar role to guilt in psychology and sin in theology. If you act blamelessly, you don't make enemies and you don't have to waste time reproaching yourself.

If he had sought guidance through the *Tao Te Ching*, Larry might have decided that he and his wife could agree to part, changing the commitment they had kept toward their marriage but remaining blameless. The forces keeping the marriage together, primarily joint parenting responsibilities and shared obligations to each other, have changed. For as long as the greater good of raising children together bound them, they tolerated the less satisfying aspects of their relationship. Understanding the mechanism of those changes helps reveal the entrance to the better path.

"When things come to the summit of their vigor,
they begin to decline. This is against Tao.
What is against Tao will soon come to an end."
—LAO TZU

I knew Larry was too well versed in strictly logical approaches for the Tao to resonate with him. But whether he heeded Kant or Lao Tzu, he might well have reached the same conclusion. According to his own strong sense of duty, Larry felt he still had an obligation to his wife and his marriage. But with his children grown, he felt he had an obligation to himself as well, and decided to leave his stagnant marriage to pursue personal growth. Kant's and Ross's ethics justified that choice. Larry was at ease with his decision once he was clear that he was acting consistently with his principles.

CARMEN

Carmen never got to make the choice of whether to stay in her marriage. After twenty-five years with her, raising four children, her husband had just left her for another woman. Carmen sought out a feminist philosophical counseling group run by one of my colleagues, Vaughana Feary. With the overarching theme of helping a client "to eliminate the intolerable, to reduce the pain, to fill the need, to actualize the dream" (in the words of philosopher Nel Noddings), Feary divides philosophical work into four phases, the first of which is for the client to tell her own story in her own words and for the counselor to assess whether referrals to any other kind of care are appropriate.

Carmen, then, unfolded this story. As her husband left, she discovered that he had been involved in a long-term affair. She had made the heartbreaking decision, not long before, to place their disabled son in a group home. These two earthshaking changes made her feel like a failure as a wife and a mother, the callings to which she had devoted her life. She'd been crying for a week, she said, not even going to church.

She hadn't consulted a lawyer, though her husband had, but she ventured that her husband would surely provide for her.

Before joining the group, Carmen had consulted a psychiatrist who diagnosed a chemical depression and prescribed Prozac. At her philosophical counselor's urging, Carmen began working with an attorney. She was then prepared to join the group and begin stage two, defining her fundamental beliefs about the good life, femininity, and feminine virtue. In Carmen's personal philosophy, good women sacrificed all for their families.

Able to think more clearly once Prozac had taken the edge off her depression, and inspired by her first task with the lawyer—estimating the economic value of her work as nurse, nanny, cook, and housekeeper (that is, housewife) over the last twenty-five years—Carmen moved easily into Feary's third phase, which entailed examining her basic philosophy for contradictions and irrational beliefs. Carmen came to see that her nurturing work—the work to which she had committed herself to the exclusion of all else—was undervalued by her husband. She ruefully realized that her confidence in her husband's not leaving her in dire economic straits was not logical, especially given how long he had deceived her in his relationship with another woman. She determined to fight for her rights in the divorce since the end of the relationship seemed inevitable. She also realized that although self-sacrifice is indeed a virtue, it alone was not enough to create a good life for her. She set out to discover what the other components might be.

Feary's fourth stage involves articulating your refined philosophy. In addition to self-sacrifice, Carmen now included autonomy as a necessary virtue. She found further evidence for this in her son's flourishing self-sufficiency in the group home, which had been impossible at home. She also came to see that each person has her own ideas about the good life and about love and happiness. She allowed, then, that although she thought her husband's abandoning the marriage was morally wrong, he might genuinely believe in the rightness of seeking his own happiness in another relationship. This insight eventually allowed her to reconnect with her other children, whom she'd been unable to face since her husband left, and resist asking them to "choose sides." Most important, acknowledging the validity of a variety of outlooks allowed Carmen to change her opinion that her life was a failure.

"The principle requires liberty of tastes and pursuits, of framing the plan of our life to suit our own character, of doing as we like, subject to such consequences as may follow, without impediment from our fellow creatures, so long as what we do does not harm them, even though they should think our conduct foolish, perverse, or wrong."
—JOHN STUART MILL

So armed, Carmen secured a favorable property settlement and alimony that reflected the monetary value of her domestic work. Though she had a degree of financial security, she exercised her autonomy by taking a job with a health insurance company that allowed her to use what she'd learned through caring for her disabled son to help families in similar situations. Eventually she went back to school for a social work degree to further her abilities to assist those families. She still held as her core beliefs the traditionally feminine virtues of empathy, nurturing, and caring for others, but she now had a wider perspective on how to apply them without losing or neglecting herself in the process. She experienced justified disappointment and anger over the end of her marriage, which she considered sacred. But she realized that although ending the relationship was not her choice, the direction the rest of her life took was her choice. She refused to wallow in the emotions aroused by a situation she could not change, and instead focused on the present moment and how to make the most of it.

JOAN

Joan came to see my colleague Harriet Chamberlain saying that her marriage was over but that she felt trapped in it because she was financially dependent on her husband. She said she was determined to go back to work—she'd left a successful career several years before to be at home with her children—and thereby open up a realistic exit from the relationship.

But as she worked with her counselor to outline options (going back to work full- or part-time, going back to school, preparing for a different field from the one she'd worked in before), Joan shot down each and every one. She didn't have the self-confidence to take a job now; she didn't have the time to attend classes; her housekeeper might resign; she had no recent work experience to put on her résumé and she'd never find a job she would want to do. No wonder she felt trapped! She found reasons why each route seemed impassable, no matter what was under consideration. Her philosophical counselor suggested that she was indeed trapped in her marriage—but not by her husband or by economic dependence. She was in a cage of her own making.

> *"To be freed from the belief that there is no freedom*
> *is indeed to be free."*
> —MARTIN BUBER

Her counselor took up existentialism with Joan because it emphasizes the recognition and actualization of one's personal freedom and responsibility for creating a life that is meaningful and fulfilling. According to Jean-Paul Sartre, confronting your existential freedom can be anxiety-producing, and to put it into practice requires awareness of the obstacles in your path—plus an understanding that you put them there yourself. The restrictions on our freedom that we put into place ourselves are called "bad faith" by Sartre.

> *"Man can will nothing unless he has first understood that he must*
> *count on no one but himself; that he is alone, abandoned on earth in*
> *the midst of his infinite responsibilities, without help, with no other*
> *aim than the one he sets himself, with no other destiny than the one*
> *he forges for himself on this earth."*
> —JEAN-PAUL SARTRE

Joan could have been a poster child for bad faith. But once her collusion in her own predicament was pointed out to her, she quickly found the courage to accept her responsibility and her freedom. With critical thinking, she began to knock down the walls she had built to keep herself in a life she wasn't happy with. As the demolition proceeded, she realized that setting herself up to go back to work but then denying every possible avenue for carrying out her plan served to let her think that remaining in her marriage was a necessary evil rather than a choice. That in turn let her deny her responsibility for staying unhappy.

Eventually Joan admitted that she actually didn't want to go back to work and that, to that end, she wanted to stay in her marriage. Once she realized that staying was her own choice, she had a renewed sense of control over her life. That allowed her to see that her self-esteem and self-confidence—the lack of which she had complained about to the counselor—hadn't been taken away from her. She had relinquished them, and so could retrieve them.

Joan began taking a more active part in the relationship, acknowledging responsibilities to it as well as obligations. Shouldering her own part in it removed some of the burden she had attributed to her husband, and she now felt she got a lot more out of it. Not only had she made her peace with staying in an imperfect marriage because she decided to do so of her own free will, but the marriage improved to the point where she valued it again.

I've included this case in a chapter on ending relationships—even though the marriage didn't dissolve in the end—to show that a relationship apparently in its death throes may be resuscitated. Things can certainly get so bad that there is no turning back (or no desire to turn back), but we are often too willing to say we've crossed that line.

Carl von Clausewitz—the most celebrated Western philosopher of war, a kind of European counterpart to Sun Tzu—wrote famously that "War is a mere continuation of policy by other means." You would be wise to remember, as you contemplate the end of a relationship, that divorce is a mere continuation of marriage by other means. There is rarely a swift, clean break. Consider carefully whether ending a relationship is going to solve or at least ameliorate the problem at hand. And if ending it is the only way to do so—or if it is ending despite your wishes—find the best way through. Once you're sure you've found it, proceed as blamelessly as possible.

8

Family Life and Strife

*"When the family is in order, all the social relationships
of mankind will be in order."*
—I CHING

*"A great proportion of the misery that wanders,
in hideous forms, around the world, is allowed to rise
from the negligence of parents. . . ."*
—MARY WOLLSTONECRAFT

The only thing more complicated than a one-on-one love relationship is the complex web of interactions in a family. Each individual
manifests his or her own personality, preferences, standards, attitudes,
values—and philosophical outlook. Nature and nurture conspire to
create an overlapping effect among family members in many of these
areas, but it is never a perfect fit. That's where the strife begins. Family
relationships need careful thought and maintenance, just as love relationships do. Perhaps they have an even greater need, because many
family relationships are inherently imbalanced and because family relationships are for the most part dictated, not chosen. You usually choose
your mate; you don't choose your relatives.

For many years, children are shaped by the adults they depend on.
That gives parents the obligation to do what they can to encourage and
instill the characteristics that will lead to living a good life. The specific
components of a good life are what philosophers have been arguing
about for thousands of years, so you're not going to find a unique blue-
print here. The details will vary from person to person, family to family, and culture to culture. The two commonalties you'll find wherever

you go are a reverence for those who strive for the good life and a mandate to parents to inculcate it, however it's defined. (There's also no preprinted lesson plan for teaching how to live a good life, so don't look for that here either.)

From the philosophical counselor's perspective, the important thing as a parent is to identify your duties and obligations and explore how you can carry them out in a way that's consistent with your philosophical outlook. Though the parental role is primary, all participants in the family structure, children included, have their own obligations to themselves and to other members of the family. What a wonderful world this would be if everyone analyzed those responsibilities and fulfilled them as often as possible.

MARGARET

During a live radio interview, I talked to a caller named Margaret who was working on this kind of analysis. She asked me about requiring her young teenage children to do basic household chores to earn their allowance. She wanted to teach her children a sense of responsibility—both in handling money and in participating in family life and upkeep. She explained that her family could certainly afford to hire someone else to mow the lawn or rake the leaves and that her kids complained that none of their friends had to work for their allowance. She wasn't questioning her decision, but she wanted to be sure there was a philosophical justification for her rule. She didn't want to exploit her children, and she didn't want them to take money for granted. When Margaret was growing up, she'd had to work for whatever pocket money she needed (in addition to her chores at home), and though it wasn't financially necessary for her kids to do the same, she wanted to instill in them the same work ethic that she'd learned from her upbringing.

To answer her children's complaints about the work, Margaret had been paraphrasing Nietzsche: a little poison can be a beneficial thing. While Nietzsche despised conventional mores like the Protestant work ethic, I thought this was a constructive application of his ideas. Doing the work would certainly not harm the children, much less kill them, despite their fervent tales of woe. And most likely it would strengthen their moral fiber by teaching them an important socioeconomic lesson: you can't get something for nothing. Or if you prefer, there's no such thing as a free lunch. That's actually an extension of one of Newton's laws of physics into the field of economics. Newton's law says that for every action, there's an equal and opposite reaction. The economics version is that for every meal, there's a check. The key questions are "Who eats?" and "Who pays?" In economics as in physics, something doesn't come from nothing. By the same token, you can't get nothing from something—it's a two-way street.

"Whatever doesn't kill me outright, makes me stronger."
—FREDERICK NIETZSCHE

Margaret had arrived at the philosophical underpinning for her actions on her own, but I threw in some further confirmation. As we'll see in Chapter 11 (on ethics), both Aristotle and Confucius saw virtue as a matter of good habits. So in Margaret's case, creating the habit of paying for one's lunch, or singing for one's supper, was helping her children practice a virtue: appreciating the value of work. Aristotle and Confucius would agree that since a virtue is a matter of habit, it can't be learned just by talking about it—it has to be practiced. Here again, Margaret was on the right track.

RITA

My colleague Alicia Juarrero worked with a client who was also struggling with her responsibilities toward her family members but in a very different way. Rita came to Alicia devastated because her teenage sister had been raped by a boy she knew from the store she worked in on weekends. This act of violence had thrown the entire family into an uproar, trying to cope with both the emotional aftermath and more practical matters like getting the girl into therapy, pursuing legal charges against the boy, and so on. Rita herself had been missing classes, neglecting her course work, and generally feeling paralyzed by the awfulness of the situation. She was at a loss for how to help her sister or her other relatives as they tried to help.

Rita was following her best instincts to love and support those closest to her. But she was in danger of losing herself in the process. That balance is often at issue in one-on-one love relationships, but it is also important within families. Both family feeling and individuality are the best way to form a healthy adult life. Her counselor pointed Rita toward the Stoic philosopher Epictetus to illuminate her situation. He wrote, "When you see someone weeping in grief . . . take care not to be carried away. . . . Do not hesitate, however, to sympathize with him." Rita was helping no one by derailing her own life. With her life in shambles, she had no resources to offer her sister. Rita resolved not to add her own stress to her sister's. Taking the time to pull herself together would be the first step toward helping her sister and the rest of her family do the same.

"If it ever happens that you turn outward to want to please another person, certainly you have lost your plan of life."
—EPICTETUS

Rita fastened onto one more insight from Epictetus for good measure: "Seek not that events should happen as you wish, but wish them to happen as they do happen and you will go on well." Something that has already happened cannot be undone, so it is fruitless to waste time wishing it could be otherwise. Better to forge ahead with circumstances as they are—no matter how distressing—than flounder in the past. Moving forward holds the only possibility for improvement.

Recall the Stoics' central theme that the only things of value are those no one can take away from you. In addition to the specific tidbits Rita gleaned from Epictetus, then, she could learn from the Stoics generally. Few things are more valuable than family love, which can never be taken away by anyone else. Even a rapist does not have that power—unless you give it away yourself. Rita was right to find a way to preserve her family's loving structure through this storm. It was more than the silver lining to the cloud they were now under; it would be the sun that relit the entire sky when the storm passed.

SONIA

Margaret was looking for philosophical justification in part because of the seeds of self-doubt planted in her by her children's resistance. Though I'm sure the children were going to lose this particular battle, they were engaging in the larger war so many families go through as children struggle to establish their identities independent of the family. I saw a much more serious case of that in Sonia.

Sonia was in her early twenties when she sought philosophical counseling. Throughout her teenage and young adult years, her mother, Isabelle, had dragged her to a seemingly endless parade of psychologists, psychiatrists, and other therapists, including a pastoral counselor. This was the first time Sonia herself had initiated any type of counseling and attended willingly. Sonia told me pointedly that she had never been diagnosed with anything. Indeed, what Sonia described to me was a basic mother-daughter conflict blown out of proportion. Isabelle, who was conservative and religious, was sure that Sonia's free-spirited, creative nature was abnormal—and could be changed if she saw just the right counselor. Sonia was convinced that none of the professionals she'd seen had helped her at all, and she didn't think her mother had been pleased with the results either.

Sonia had been a rebellious teenager and even as a child felt that she didn't fit in at school or at home. Isabelle had long seen every example of Sonia's exerting her own will as deliberate misbehavior and as a sign that something was seriously wrong with Sonia. Eventually Sonia herself began to fear she was somehow abnormal.

Sonia worked as a model, attended college part-time, and lived with her parents (her most affordable option). Isabelle objected to Sonia's work and her studies. Modeling was sinful, according to Isabelle's religious beliefs, and she didn't think a degree in art history was a worthy investment of time or money. Sonia's father remained in the background, overshadowed by the women in the house, and neither supported nor opposed Sonia's or Isabelle's actions toward each other.

Sonia sought me out because she wanted to resolve for herself whether there was something wrong with her and whether her choices about how to live her life were immoral, as Isabelle said. In her heart Sonia didn't believe those things, but the conflict with her mother had been so long-running that she had nagging doubts. She was really looking to establish her own identity. What kind of person was she? What were her own standards? Were her standards as good as anyone else's?

Sonia and Isabelle were engaged in one of philosophy's traditional battles: relativism versus absolutism. Relativists hold that principles and actions are not intrinsically right or wrong but that cultures and individuals assign them values (e.g., beauty is in the eye of the beholder). In this way of thinking, no one thing is inherently better or worse than any other thing. Our aesthetic and moral precepts are up to us; there is no objective way to judge them. To Sonia, this way of looking at the world made the most sense. She respected her mother's religious views, and although she chose not to follow them herself, she also didn't ask her mother to abandon them. Isabelle, on the other hand, was an absolutist, with a more clear-cut, black-and-white worldview. To her, some things were right and some things were wrong—no ifs, ands, or buts.

"Fire burns both in Hellas and in Persia; but men's ideas of right and wrong vary from place to place."
—ARISTOTLE

"Man is the measure of all things."
—PROTAGORAS

Theoretically a relativist should be able to get along with an absolutist by acknowledging that absolutism is as valid a way of looking at the world as any other. An absolutist is going to have more trouble with a relativist, and this seemed to be just the situation Sonia and Isabelle were in. Sonia was tolerant of her mother and wanted nothing more from Isabelle than the same kind of acceptance of her choices.

For the last thirty years, relativism has dominated Western thought. Like anything else applied absolutely, relativism has its problems, both logically and practically. If you think relativism is definitely the best way to look at things, then you're already encroaching on absolutism— or at least absolute relativism. Try asking relativists if murder, or rape, or slavery, or genocide is morally permissible. Most of them will say no—and then you've caught them assigning something an objective moral value. (You can then restore order to their universe by discussing self-defense, or abortion, or capital punishment and the relative morality of different types of immorality under different circumstances.)

It doesn't take long for relativism to become self-contradictory. There's a famous story in academic philosophical circles about a professor facing a class full of self-proclaimed relativists. Following several heated class sessions during which the students denounced absolutism of every kind, the professor gave them all Fs on their end-of-unit essays. When protests erupted, he explained that they had convinced him that everything was relative and therefore subjective, and in his subjective opinion all their essays were worthless. Soon his office hours were booked solid with former relativists now arguing that their work was objectively good—and better than others'—and demanding improved grades. Relativism is fine until it costs you more than you're absolutely willing to pay.

In real life, textbook relativism doesn't work. To stave off anarchy, society must regulate acceptable conduct at some point. Most people in decent societies agree on a list of forbidden things, including murder, rape, incest, and theft. If you allow for a limited set of objective values, however, a subjective outlook can still work in many other matters. From my perspective, which is close to Mill's, some things should be relative—as long as they don't bring harm to other people or infringe on other people's liberties. Certainly in this case I agreed that Sonia had the right to expect her mother to respect her individual integrity as an adult and allow her to make her own choices. They each had the liberty to think whatever they liked, but not to impose conformity on the other.

To be able to enjoy their mother-daughter relationship—or at least to calm it as long as they were living in the same house—Sonia and Isabelle needed to reach an accommodation with each other. Since Isabelle wasn't seeking counseling and seemed unlikely to change her ways, Sonia focused on what she could do herself. As she accepted her propensity to be herself, knowing that she was normal, Sonia gave up feeling assaulted by others telling her she was different and asking her to change. In becoming comfortable with herself and seeing her own standards and values as valid, she stopped rebelling as a matter of course—without prodding from her mother, or her counselor, or anyone else. There were fewer explosions at home, and Sonia's college grades even improved. When Sonia's behavior naturally changed, her mother stopped giving her such a hard time. Isabelle realized that having different values does not equal "mental illness."

With the air clearing up, Sonia was able to tell Isabelle, "I am who I am. If you get to know me, you might even like who I am, or at least parts of me." Philosophical counseling allowed her to believe in herself on her own terms and encourage her mother to do likewise. Sonia was also prepared for her mother to refuse to do so. Over the year or so that I counseled Sonia, she and her mother came to agree that one of them could spend Saturday night at a church service while the other spent it in a nightclub, and they could still respect each other in the morning.

Though it was Sonia who sought philosophical counseling, this issue, like most problems in any relationship, was the product of more than one person's actions. If Isabelle had been the one making a weekly appointment with me, I would have discussed an overlapping set of philosophical ideas. The importance of living in accord with your own worldview and the issues surrounding relativism would of course still apply. But for a parent there are additional responsibilities regarding a child. The younger the children, the more the burden rests on the adult, but the load becomes more evenly shared as the children themselves become adults. (Later in life, roles may become reversed, as we'll see in the case of John later in this chapter, and the grown child carries much of the responsibility for the aging parent.)

Humans need the love of a family, biological or otherwise, to grow up feeling worthy and secure. Parents, and other adults caring for children, have a responsibility to their children—and to society—to provide that love. Part of a parent's job is to inculcate virtues (good habits, according to Aristotle). But everyone gets to a point where they need to reinforce their own self-worth, live by their own values, and take their own independent adult place in society. Parents have a duty to prepare children by helping them lead lives of integrity. But once your duty is done, you can stop doing it. Part of your duty is to stop before parental authority becomes parental intrusion. Fulfillment comes from inside oneself, an Aristotelian idea introduced in Chapter 5. No one else, not even a parent, can give it to you. What the best parents can do to encourage ultimate fulfillment for their children is to allow them room to become self-reliant as they grow.

Children are extensions of their parents biologically but not always culturally. Parents have a genetic investment in and legal custody of their children, but they do not have proprietorship. Kant frames the issue as treating all people, including children, as ends in themselves, not a means to your ends, as we've discussed in earlier chapters. The Lebanese poet and philosopher Khalil Gibran devotes a beautiful passage to raising children, which centers on this idea of parents having temporary stewardship but not ownership.

"Your children are not your children. They are the sons and daughters of Life's longing for itself. They come through you but not from you, and though they are with you yet they belong not to you. You may give them your love but not your thoughts, for they have their own thoughts. You may house their bodies but not their souls,

for their souls dwell in the house of tomorrow, which you cannot visit, not even in your dreams. You may strive to be like them, but seek not to make them like you."
—KHALIL GIBRAN

Children begin to forge their individual identities in the crucible of the family. The more divergent those identities turn out to be from those of the parents or the family as a whole, the more potential there is for conflict, as happened with Sonia and Isabelle. We humans depend on our parents for a much longer period than any other mammal because we mature slowly, taking many years to learn all we need to know to assume a fully adult role in society. We mature by degrees, which our laws reflect by staggering the ages at which it becomes legal to give consent, vote, drive, drink alcohol, marry, join the armed forces, and so on. The more a child takes on, the more parental control lessens. The greatest potential for conflict occurs at the moment when the child's opinion counts as much as the parent's. That full autonomy comes with adulthood, and the long and winding road leading to it accounts for much of the chaos of adolescence: the child hungering for autonomy, the parent struggling with letting go, and reality not yet (or no longer) eye-to-eye with either party's desires.

Sonia and Isabelle were reaching peace at an important time for them; shortly thereafter Isabelle was diagnosed with cancer. Both mother and daughter then faced a new kind of philosophical struggle, which we will discuss more in Chapter 13. Part of what happened after the diagnosis (and could never have occurred when the two women were so embattled) was that Sonia stepped in as her mother's primary caretaker and lovingly ministered to her. This kind of role reversal of parent and child is becoming more and more common, and it presents a whole new set of philosophical issues, as we'll see in John's case.

JOHN

John's mother, Celeste, suffered from a degenerative neurological disorder and was confined to a wheelchair. John lived at home with her, in part because of his low income as a graduate student and in part because of the care Celeste needed to be able to live in her own home. But during the past year, Celeste's bouts of disorientation and anxiety had increased in frequency, so John worried each time he left her alone to go to school or work. Because she was no longer always lucid, Celeste had assigned medical and legal power of attorney to John.

Then the kind of thing John had feared actually happened. He returned from an outing and found his mother at the bottom of the staircase, unconscious and bleeding, having tried to negotiate the stairs in her wheelchair. At the hospital, the doctors found no serious injuries but admitted her for observation and urged John to place her in a nursing home. They felt that despite John's attentive care, Celeste had come to a point where she needed constant supervision. The team of doctors caring for her, together with the hospital social worker, suggested that despite her lack of injuries, Celeste stay in the hospital until a place became available—which could take months.

John accepted the fact that Celeste would eventually need more care than she could receive at home, but he feared that both her state of mind and her general physical condition would deteriorate in the hospital since she didn't need medical treatment and would have very little stimulation on a medical ward. Celeste, when she was lucid, said she wanted to go home. John was willing to do whatever he could to allow his mother to live at home, at least until a place became available in a good nursing home, but there would still be periods of time when he had to be away from her, and he didn't have enough money to hire an attendant. He envisioned spending a last summer at home with his mother before she had to move. He wanted to grant her wish to go home but also felt it would be easier to talk to her about a nursing home—which she would no doubt resist—if they could do it in their own time, in their own home.

John brought his dilemma—whether or not to leave his mother in the hospital while waiting for nursing home care—to me because he felt unsatisfied by either alternative and wanted to clarify the ethical implications of both. He actually needed to address two different aspects of philosophy, ethics and decision-making. He faced the ethical question of what it means to be responsible for someone else's well-being and in whose interest you should be acting—theirs or yours—and when. John faced this question as a child caring for a parent, but this kind of question even more frequently applies to parents caring for young children. Secondly, John needed to explore ways of making a decision and how to choose between two difficult options so that he would feel ethically justified in the course he took.

Like many other clients, John had worked his way through the first three stages of the PEACE process. He obviously understood the problem, was taking both his and his mother's emotions into account, and had done some useful analysis in identifying his two options and their possible consequences. But this was not enough for John to work with. He needed contemplation to cultivate a disposition that would allow him to make a tough choice one way or the other.

I used decision theory to help John through the contemplative phase. Decision theory is a philosophical name for the mathematical theory of games founded by John von Neumann and Oskar Morgenstern. This theory uses games metaphorically to encompass many human activities where participants decide on the best move to make under some set of rules but usually without all the facts on the table. Decision theory, as used by philosophers, captures the main ideas of game theory but usually avoids extreme mathematical complexities.

Only in a small subset of games is there actually a best move at every given stage. In such games it is rational to make that best move if you can find it. But in most games there is no best move at all; different strategies point to different choices. Then the question is not simply "What move is it rational to make?" but rather "What strategy do I prefer to adopt?"

> *"The importance of the social phenomena, the wealth and multiplicity of their manifestations, and the complexity of their structure, are at least equal to those in physics. . . . But it may safely be stated that there exists, at present, no satisfactory treatment of the question of rational behavior."*
> —JOHN VON NEUMANN AND OSKAR MORGENSTERN

If John were playing chess, or tic-tac-toe, he would be looking for his single best move on each turn. Those games fall into the limited category of strictly determined games, defined by being for two players, being zero-sum games (winnings balance losses), and being games of perfect information (nothing is concealed, all the moves are on the table). In a strictly determined game, there is always a best move, and all you have to do is find it. If you do find it on every turn, you can't lose—the worst that can happen is a draw. The best move is easier to find in tic-tac-toe than chess, of course, but the principle is exactly the same. Children get bored by tic-tac-toe once they discover they can always at least reach a draw, and chess masters often agree to call a game a draw after only a few moves. Although in some strictly determined games it can be extraordinarily difficult to determine your best move, at least you know there is one to be found.

Life, however, is not a strictly determined game. Unfortunately for John, his dilemma regarding Celeste was a more common, and less prescriptive, kind of game. There are more than two players, once you take into account the doctors, nurses, social worker, and others who may be involved in the case. It isn't a zero-sum game, since the potential losses (injury or death) are not equivalent to the potential winnings (time together). It isn't as clear-cut as poker, where if you lose $5, one or more players win that $5. And it certainly isn't a game of perfect information—no one knew exactly how or when Celeste's illness would manifest next. But decision theory can still be useful in representing such a game in terms of one's choices and their possible consequences, to give one a clearer picture of the situation. John's decision matrix looked something like this:

TABLE 8.1

	Possible Consequences	
John's Choices	Best Outcome:	Worst Outcome:
bring mother home:	wonderful summer together	serious or fatal accident
leave mother in hospital:	medical supervision	psychological deterioration

The decision matrix illustrates that there is no best choice according to both possible outcomes. While having a wonderful last summer together at home would be far better than just medical supervision on a hospital ward, a serious or fatal accident at home would be far worse than becoming bedridden and deteriorating psychologically on a hospital ward.

Decision theory does not tell you how to play, but it can help establish what kind of criteria you should be using in deciding your moves. You must understand the nature of the game to make a choice. If you know there's a best move, you must try to find it. In other cases, such as John's, you must ask yourself: What do I want to gain? What do I want to avoid? What am I willing to risk? What are the other players aiming to gain and avoid, and what will they risk? Short of being assured of the existence of a best move, following decision theory might mean estimating the odds of each possible outcome occurring, weighting the likely benefits and drawbacks, and choosing the path most likely to produce the most benefits.

So if John elects to bring his mother home, he is essentially playing a game of chance without knowing the odds. When you play a casino game, at least you can calculate the probabilities. If you do try to set odds, steer clear of two versions of the "gambler's fallacy." Version one says that the gamble with highest payoff is best. That disregards the odds entirely, so is often not a safe bet. In John's case, this version of the gambler's fallacy would prescribe bringing his mother home, since that choice entailed the possibility of the best payoff. But it also entailed the possibility of the worst!

Version two of the gambler's fallacy states that whatever just happened is unlikely to happen again. If you toss a coin five times in a row and get five heads, the fallacy states that tails becomes more and more likely on each subsequent throw. This is false because the coin has no memory; each toss is an independent event. There is no such thing as red being hot on the roulette wheel—the wheel has no memory, and (unless you're in a crooked casino) each spin starts with the same probability of any given outcome. Rolling snake eyes twice in a row makes it no more or less likely that the next roll will be—or will not be—snake eyes.

John's mother was suffering from an illness that left her intermittently disoriented. We couldn't estimate the odds that she would become disoriented again in the next hour, or day, or week, and we couldn't know exactly what position she would be in (e.g., about to go downstairs) the next time she became disoriented. So John could assume neither that because his mother had just fallen she would be injured again at home, nor that since she had just been hurt, she'd already had her share of misfortune and would remain safe at home.

I agreed with John that he was in a very difficult spot, since both options entailed advantages and disadvantages, both for himself and for his mother. John and I laid out both best-case and worst-case scenarios. He imagined his greater pain and guilt, and her additional suffering, if she sustained a severe or even fatal injury while left alone at home. In the hospital, John imagined her facing physical and psychological deterioration that would make the transition to a nursing home all the harder. From this worst-case analysis, John concluded that both he and his mother would be better off if she remained in the hospital. A best-case analysis, however, led to the other conclusion. At home, mother and son could enjoy a shared summer and prepare for the next stage. In the best of all worlds, both John and Celeste would be better off with her at home.

It was not my job to recommend a choice between bringing her home and leaving her in the hospital. But I did outline the nature of moral responsibility when making a decision on behalf of another person. The responsible path is to decide what is best overall for the other person, not what is best only for yourself. We have this kind of responsibility for our children, and also for our infirm parents.

We also have to allow others to choose within the limits of their autonomy. If you decide it's OK for your child to have an ice cream, surely your child can choose the flavor. Even someone about to be executed by the state can choose a last meal. Since John had already decided to place his mother in a nursing home, maybe she could choose how to spend her remaining months outside the home—notwithstanding the risk.

What is best can mean whatever helps that person avoid the worst; it can also mean whatever helps that person attain the most. The key in deciding for another is to set one's personal gains and losses aside.

After two sessions, John told me that he could see the decision-theory and ethical boundaries he was operating within. He said he would be able to arrive at a decision that was justifiable to him. John had neared the end of the contemplation stage and would make his decision in equilibrium. I do not know—nor do I need to know—what John decided to do. As a philosophical counselor, my responsibility is to help my clients attain philosophical self-sufficiency, not dependency. Instead of agonizing over a difficult decision, or wallowing uselessly in an emotional quagmire, or being diagnosed with a bogus personality disorder, John could occupy philosophical high ground. He could feel the sadness of the situation but could get beyond indecision. Sometimes we should feel sad—and there is even a kind of solemn joyousness concealed in that—but we should never have to feel indefinitely incapacitated by sadness.

"The best thing for being sad is to learn something. . . . Learn why the world wags and what wags it."
—T. H. WHITE

9

When Work Doesn't Work

---

*"The return from your work must be the satisfaction which
that work brings you and the world's need of that work.
With this, life is heaven, or as near heaven as you can get.
Without this—with work which you despise, which bores you,
and which the world does not need—this life is hell."*
—W. E. B. Du Bois

*"Work keeps us from three great evils:
boredom, vice, and poverty."*
—Voltaire

Work is a large slice of life in general, so many workplace issues overlap with other topics of this book. Many of the work-related issues my clients bring me are interpersonal issues at heart and some of Chapter 6, on maintaining relationships, applies. For that matter, so do some considerations from the chapters on seeking and ending relationships and on family life. Some clients are trying to resolve ethical conflicts that arise at work or to explore the moral implications of managing others. We'll look at ethics and morals in Chapter 11. Some are struggling with questions of the meaning or purpose of their work (see also Chapter 12), doing fulfilling work, and establishing a balance between work and the rest of life. But since most of us spend more time working than doing anything else, it's important to consider work-specific issues in their own right.

Doing a job well gives us a sense of fulfillment, no matter what the job is. Most people want to do a good job and want to be praised for doing a good job. If you're the boss, take note: acknowledging your subordinates' achievements is a true motivator. But if you have had a long wait for justly deserved praise, try to take satisfaction just from knowing you did a good job. Your desire to be praised is natural, but if the praise isn't forthcoming, there's nothing to gain from dwelling on it but unhappiness. The *Bhagavad Gita* underlines the importance of doing good work for its own sake.

> *"Let not then the fruit of thy action be thy motive;*
> *nor yet be thou attached to inaction."*
> —BHAGAVAD GITA

The *Bhagavad Gita* is a widely translated Sanskrit poem relating a dialogue between a warrior prince (Arjuna) and a human incarnation (Krishna) of the god Vishnu. On the eve of battle, they debate the ethics of fighting and killing (or being killed) and the nature of duty. That it is military advice might seem to lend it obvious relevance to today's corporate battlefield, and the ultimate message of selfless dedication to a higher power might please generals both civilian and military. But this book hasn't been widely read for nearly three thousand years because it urges corporate loyalty. The driving force is the value of performing a duty for its own sake and in the service of a higher principle, rather than working merely for paychecks and perks.

We have all encountered workers—whether in public service or the private sector—who seem to be interested primarily in watching the clock, longing for Friday, or waiting for payday. They are not interested in their work, only in its fruits. By being attached primarily to the fruits, they impoverish their labors. By impoverishing their labors, they displease those whom they serve—both employer and client—which further impoverishes their labors. By this vicious circle, they diminish the fruits themselves. By contrast, we have all encountered workers who perform their tasks primarily in the spirit of service and who seem moreover to love what they are doing. This devotion enriches their labors, which pleases those whom they serve—both employer and client—which further enriches their labors. By this virtuous circle, they increase the fruits of their labors. They do so by not focusing solely on the fruits themselves.

Most of us appreciate works of art such as poetry, painting, and music, and cultured societies reserve some of their highest praise for great artists. In the act of creation, poets, painters, and composers are fully absorbed in the labor of bringing forth their art, not on its fruits. If you do your work well, the fruits ripen themselves. If you fantasize about tasting the fruits instead of working well, they won't ripen at all. You too have the power to make your job a work of art. Vow to be like a great artist at whatever you do.

"If you work at that which is before you, following right reason seriously, vigorously, calmly, without allowing anything else to distract you, but keeping your divine part pure, as if you were bound to give it back immediately; if you hold to this, expecting nothing, but satisfied to live now according to nature, speaking heroic truth in every word which you utter, you will live happy. And there is no man able to prevent this."
—MARCUS AURELIUS

COMPETITION

Doing a job well, from a philosophical perspective, doesn't necessarily mean doing it to perfection or doing it better than everyone else. There's no moral significance to winning or losing a race. The winner may be the fastest runner, but that has no bearing on whether she is a good person. The value lies in working hard and doing your best. It may be that your best doesn't get you across the finish line first—or get you the corner office or a big raise—but if you have striven to do as well as you are able, you have earned personal satisfaction. The Stoics have pointed out that satisfaction is what is valuable about work: the result no other person can take away from you; the part no one else has power over but you.

The key is how you measure your best. Ours is a competitive culture, and we are competitive by nature. Using the performance of others as your only yardstick is a mistake. Not using it would also be a mistake. Competition brings out the best as well as the worst in individuals. Go jogging with your quicker neighbor and see if your time or endurance doesn't improve. At the other end of the scale, competing graduate students have been known to rip pages out of weekly scientific journals to make it more difficult for their peers to learn about the latest developments. Competition isn't by definition bad, but it can be a destructive force.

As much as our society rewards particularly aggressive competitive behavior (as in professional sports), some sectors have become terrified by even its mildest forms. If your child's school even has a field day anymore, I'll bet it's rigged so that every kid wins a ribbon. Maybe even every kid entered in an event gets a prize. If the idea is to build self-esteem, this strategy of empty gestures has the reverse effect. If everyone gets a ribbon, why run?

It's only natural that some racers are faster and others slower. We should recognize the fastest runners if we value swiftness. But we should not confuse fleetness of foot with excellence of character. John may be a better runner than Jack, but that doesn't make John better than Jack. Creative, constructive competition allows you to discover and express your abilities. At work, the trick is to find a balance between competition and cooperation.

MEANINGFUL WORK

Work is a journey for most people. Few of us are born to do one thing in particular. The way you achieve greatness in your life is specific to you; there is no foolproof recipe. Most people do not feel called to their work, but finding the right work for you is one of the surest ways to achieve fulfillment. We all have particular talents, but most of us have to dig a bit to find ours. Finding them—and determining how to use them—contributes a sense of meaning or purpose to daily life. It doesn't matter how high or humble your aspirations are. The principle remains the same whether you're a Fortune 500 CEO, a stay-at-home parent, a volunteer with Mother Teresa's organization in Calcutta, a janitor, a sculptor, or an office mate of Dilbert's. Meaningful work is integral to a meaningful life. Being laid off is so painful because of attachment to position, status, privilege, and security, not just general financial straits. Retirement is often so difficult for the same reasons.

> *"All work, even cotton spinning, is noble; work is alone noble. . . . A life of ease is not for any man, nor for any god."*
> —THOMAS CARLYLE

Dilemmas involving fulfillment are another issue. For instance, fulfillment can come from parenting or from a career. It is tempting but difficult to balance both. Many parents are torn between work outside the home and caring for their children full-time. Some find full-time parenting completely fulfilling, while others aspire to climb a career ladder. Many more find fulfillment in both places but are conflicted about how to do both jobs well.

How Will You Know Unless You Try?

Most people will hold a variety of jobs over the course of their working lives, and more and more commonly those jobs aren't even in the same fields. We've come a long way from the days of the company man, signing on with a corporation as soon as he finished school and leaving forty years later with a gold watch and a handshake to attest to his loyal service. Some people change course because they are forced out by their companies. Some people move on because they find greener pastures and don't feel any more loyalty to their company than they know their company feels to them. Others burn out or simply want a change of pace. Many are simply looking for the path that will lead them to fulfillment. They are trying on jobs or fields to find the ideal fit.

Working at a job you don't like isn't necessarily bad. Sometimes you have to do a job to discover you're not suited to it. You can't know everything through reason alone, as the rationalists would have it. Some things you need to learn through experience. Experience isn't the only teacher, however. We also need to reason about our experiences. There, in the Golden Mean between strict rationalism and strict empiricism, lies a sensible path for living your life and learning from it as you go. You can learn valuable lessons even from a bad experience. If you're seriously contemplating whether some job opportunity is good for you, there's only one way to find out for sure: try it.

> *"The only way one can learn to recognize and avoid the pitfalls of reflection is to become acquainted with them in application, even at the risk of gaining wisdom by sad experience. It is useless to preface philosophizing proper with an introductory course in logic in the hope of thus saving the novice from the risk of taking the wrong path."*
> —Leonard Nelson

That great philosopher for children, Dr. Seuss, expounds on this theme in his famous treatise *Green Eggs and Ham*. The opening empirical question, "Do you like green eggs and ham?" goes unanswered until the closing pages because our hero refuses to try them. When he finally agrees to a taste, he reverses his earlier adamancy on the topic and declares, "I do so like green eggs and ham!" This can cut both ways, of course. After a night out with the boys—some very experienced party animals—Voltaire was invited to join them again the next night. He declined, explaining, "Once, a philosopher; twice, a pervert."

You may never know unless you try. But unless you reflect on your experiences, you won't be able to enlist them in your progress. You can save yourself time and trouble by using your reason and experience to choose a likely avenue for yourself. But if your best efforts still lead to a disappointing, unfulfilling, unchallenging, impossible, or otherwise unhappy career move, know that you haven't necessarily wasted your time—as long as you file what you've learned and make use of it next time.

Conflict

Any time people come together, there will be personal differences and conflict. That's true for athletic teams, political parties, academic committees, religious orders, and office staffs. In fact, since you usually don't choose the people you work with, extra effort may be required to function well, just as with families. On the plus side, the shared purpose engendered by working toward the same goals should help keep the wheels turning. As we learned from Hobbes in Chapter 6 (on maintaining relationships), having an outside authority you've all agreed to defer to is a key component of maintaining peace. That could mean simply reporting through the same chain of command, but for best results a sense of higher duty is required. Everyone having to "cc: Mr. Smith" on all memos shares one kind of bond, but the stronger connection comes from being on a mission together to build a house or educate a child or save the wetlands or meet the sales quota. With the kind of awe-inspiring external power Hobbes wrote about, you can avoid war. But don't count on harmony. Having realistic, not idealistic, expectations helps maintain a philosophical outlook when problems do arise.

Jean-Jacques Rousseau argued that people are basically good but corrupted (into political beings, among other things) by civilization. In a state of nature, he says, we wouldn't be so inclined: that's his idea of "the noble savage." Rousseau and his fellow romantics were rebelling against authoritarian society, and it might be tempting to believe we'd be better people if only we didn't have to live in a civilized world. Americans are conducting Rousseau's experiment wholesale these days: we've got incivility aplenty. Does that make people better? If today's postmodern descendants of Rousseau succeed in transforming Americans into illiterate, acultural barbarians—and they have already made great strides—we will find out just how misguided Rousseau was. His path leads not to Eden but to anarchy.

Aristotle argued that humans are innately political animals, and if he is right, as I believe he is, there will always be politics, including office politics—which probably used to be cave politics. The profitable debate is not whether we are political animals; it is about what kind of civilization offers its citizens better lives. No civilization at all gets Rousseau's vote, but there are other candidates.

"Man is born free, but everywhere he is in chains."
—JEAN-JACQUES ROUSSEAU

"Man is by nature a political animal."
—ARISTOTLE

VERONICA

Veronica had the kind of job newly minted journalism majors would kill for. She researched, produced, and wrote stories that were broadcast to a national audience. The catch: her stories went out fronted by the major journalism star she worked for. This big star had a famous name, a large following, a sizable paycheck, and probably all the other things that routinely accompany such a career—including some major headaches. He also had a first-rate assistant who made all that possible.

But Veronica herself felt invisible. Nobody knew her name, she didn't even have a boyfriend, let alone a fan club, and her paycheck, though certainly a tidy sum, was paltry compared with what *Variety* reported her boss had just re-signed for. Veronica felt she deserved much more credit for what she did.

Let's look at Veronica's case through the lens of the PEACE process. Her problem, she said, was that as hard as she worked, she felt unfulfilled. Her emotions were also clear to her: frustration, anger, dissatisfaction, envy, loneliness. As far as she'd gotten with analysis, she saw few options. She often fantasized about telling off the big star as she handed in a resignation letter—just before slamming the door—but she had the kind of coveted position anyone serious about a career in journalism couldn't just walk away from.

At the contemplation stage, I asked Veronica to consider whether her unhappiness stemmed not from unfulfillment but from attachment—not to the fruits of her own labors but to the fruits of someone else's. She was proud of the work she did, which is why she wanted more recognition for it, and she loved the process. In most ways, she found her work satisfying—not exactly the hallmark of unfulfillment. But her attachment to the circumstances of her boss's work obstructed her own sense of fulfillment, which must come from within, not from outside. If she could rid herself of her attachment, she would rid herself of her unhappiness. By asserting that something out of reach was "to die for," she was already dead. That is, she didn't see the splendor of what she had now, who she already was, and all there was to live for.

> *"Renounce the craving for the past, renounce the craving for the future, renounce the craving of what is in between, and cross to the opposite shore."*
> —BUDDHA

> *"It is better to do one's own duty, however defective it may be, than to follow the duty of another, however well one may perform it."*
> —BHAGAVAD GITA

Hindu and Buddhist thought made sense to Veronica, and as she built them into her own philosophy, she revisited her analysis of her situation. Maybe it was true that the big star didn't value Veronica as much as she'd like. But that shouldn't make her incapable of valuing herself. If he couldn't thank her personally once in a while or find some way to show his appreciation, then he had a problem: either blindness or ingratitude. But that was his problem, not Veronica's. Still, Veronica needed to realize that she was essential to the big star's success—but not irreplaceable. Remember those thousands of new graduates drooling for her job?

Since Veronica already knew she was fortunate to have her job, she wanted to want to keep it. Her new disposition—nonattachment to the "stuff" that comes with a high-profile job and finding personal value in a job well done—helped her reach equilibrium. Buddhist philosophy asserts that outward conflicts (between people) are almost always a product of inner conflicts (within people). As Veronica resolved her own internal issues about prestige and recognition, the conflict between her and her boss dissipated and she recaptured her original enthusiasm for the importance and excitement of her work.

TEAMWORK

Veronica was part of a team. But she wasn't like a batgirl, who hands the bat to the big slugger (that's nonessential: he could always fetch his own bats). She was more like the catcher calling the pitches (that's essential: the pitcher's lone judgment isn't as strong). A team is always more important than its individual players. Players come and go; the team remains. If you truly love the game and sincerely serve it, you'll be happy to play your position, whatever it is. If you love the game but think you're playing the wrong position, talk to the coach or try playing for a different team.

Very little gets done without teamwork, and on any team there are naturally leaders and followers. That's a good thing: imagine an orchestra sans conductor or a football team without a coach and a quarterback. Or worse, a team made up only of quarterbacks, or an orchestra where each chair had a baton as well as an instrument. Hobbes was a fan of recognized authority, as it serves both to keep the peace and to get things accomplished.

From an evolutionary perspective, humans are genetically programmed for hunting and gathering. To gather, you just forage around in a likely place, without a major plan or a lot of coordination. Gathering is neither highly cooperative nor highly competitive. But hunting is different. To hunt well, a unifying plan is the key to getting the job done. The hunt was originally a collaborative effort, something done in bands to increase the likelihood of success at a time when successful hunting was critical for survival. The results were shared among participants. Even in that communal setup, however, one individual, or a small subgroup, would have been in charge of organizing the band, mapping out a route, appointing scouts, and whatever else had to be done to up the odds of restocking the larder. Even if those with lesser talents for some reason held positions of authority, the success of the hunt would depend on following the lead of whoever everyone agreed was in charge. Otherwise everyone would be crashing through the woods in all directions, scaring away the quarry and worrying about hitting one of their dispersed companions rather than taking confident aim. You might prefer to go out on your own, but if you weren't successful, you couldn't always expect a share of the others' evening meal.

DIFFICULT BOSSES

One of the work complaints I hear most frequently is about difficult bosses. I'm sure that pattern goes way back, and there were plenty of complaints about having to listen to a guy who couldn't find a deer if it walked up and introduced itself, or getting yelled at every time you missed a shot and then again when you hit the target but didn't make a bull's-eye. If you are working under a boss you can't stand, one option is to find a different job. Of course, there's no guarantee your new boss will be better, and there's no guarantee you'd be able to keep working for the same boss if you did get one you liked.

If the only flaw in your current situation is the person you report to, think carefully about all aspects of your work before you decide to move. A more adaptable choice would be to take a philosophical attitude about working with such a person that allows you to rise above the messiness. You may have to absorb unfairness, and developing your own inner resources will help you cultivate a thicker skin. Find someone else in your company or field to mentor you, to provide some positive input to balance working for a difficult person. Remember to ask yourself, "What can I learn from this?" The answer may make the experience worthwhile.

Difficult though it may be, this is probably the "best way" the Tao recommends. If you've found that path, the Tao teaches that no one can hurt you. It's not just a question of being nice so that no one wants to stab you in the back. On the highest path, your back simply isn't available.

> *"For I have heard that he who knows well how to conserve life,*
> *when traveling on land, does not meet the rhinoceros or the tiger;*
> *when going to a battle he is not attacked by arms and weapons.*
> *The rhinoceros can find nowhere to drive his horn; the tiger can*
> *find nowhere to put his claws; the weapons can find nowhere to*
> *thrust their blades."*
> —Lao Tzu

As we saw in Chapter 6 on maintaining relationships, Hegel's theory about the master-slave relationship explains that slaves in fact hold some power over their masters. Masters are dependent on slaves emotionally as well as economically. In the earlier chapter, I used Hegel's ideas to advocate a balance of power in a romantic relationship to avoid the warped dynamic of master and slave. But a boss and subordinate are, by definition, in a hierarchical relationship, so a balance of power should not be your goal in a work setting. Rather, in this context, Hegel's ideas should help you realize your own power in the situation. That may not fix it, but it should make you feel better.

Just because your boss tells you that you aren't doing a good job doesn't mean you aren't, and you probably don't have to fear for your position. Your boss needs you. If nothing else, the pecking order requires someone to be pecked. Maybe your boss is yelling at you because her boss just yelled at her. So don't yell back, don't take it personally, and don't go home and take it out on your dog. Do your work to the best of your ability, and take the high road of letting the vicious cycle end with you.

If you are a boss and want to know that your subordinates will do their best work for you and not burn you in effigy behind your back, philosophy has lessons for you too. (If you simply want no-questions-asked compliance, there's also some philosophy to guide you. See Machiavelli: "Since love and fear can hardly exist together, if we must chose between them, it is far safer to be feared than loved.") Kant formulated the central idea: treat others as ends in themselves, not means to your own ends. You can still get someone to write up the report you need, but you should recognize whoever does it as a human being, not a report-writing machine. That's the distinction Buber made in I-It versus I-Thou relationships. You should have a different relationship with a machine that spits out a report (an It) and a person who does (a Thou).

Lao Tzu advised leaders to demonstrate humanity, compassion, and mercy as signs of strength and advised, "Govern a great state as you would cook a small fish." For those who don't know their way around a kitchen, that means do it gently. In the West we've been trained to be tough and decisive, to look upon gentleness as the soft underbelly—where you're vulnerable. The Tao, on the other hand, teaches that the true sign of strength is that you can afford to be gentle. The Golden Rule (which nearly every religion has a version of) also applies: Do unto others . . . To be respected as a boss, you must respect your employees.

> *"The best employer of men keeps himself below them."*
> —LAO TZU

A GOOD RAILROAD BOSS KNOWS
HOW TO DRIVE A TRAIN

Because she looks at the world through a capitalist lens, Ayn Rand has perhaps written more about all aspects of work than any other philosopher. Her vision of the ideal boss is reflected in her characters. Her heroes and heroines work their way up and bring the knowledge of where they've been into the highest offices with them. In a Rand novel, even the child of the owner works at the steel mill, learning to do all the jobs there in preparation for taking over the reins. Another character, a woman who owns a railroad, also knows how to drive a train. She has mastered the pieces of the business and knows how to put them together. Rand acknowledges that not everyone who can learn to drive a train will also be able to run the whole railroad. But to run the railroad, she believes you must know how to drive the train. To Rand, the boss must not be above sweeping the floors. There is nothing menial about any work; no job that needs to be done is beneath anyone's dignity.

Rand's point is that the most effective leadership is leadership by example. Senators who have been inside the Beltway too long lose their sway in their home states. You inspire respect not by title alone but by knowing what needs to be done and being willing to pitch in. The most vehement criticism leveled at those who have risen through the ranks is that they forget where they came from. Forgetting is a major misstep: not only could you use your knowledge to do your current job better, but it could also win the authentic respect of those who work under you. And most importantly, that knowledge will never let you forget that we are all in it together as human beings, no matter what titles appear on our business cards.

Most of us have lost touch with all other work save our little piece of our chosen field. We don't know about other people's jobs or other people's labor. Yet we depend on the work of others for all our needs. Do you ever stop to think where those tomatoes in the supermarket in February come from? That disconnection is to be expected in a highly technological global society, starting from the most basic point that what we consume is increasingly unrelated to what we produce.

Zen Buddhism offers another perspective on doing basic work. Zen teaches that routine work is valuable in and of itself. Humbling yourself is a path to self-improvement in the Buddhist tradition. Any task undertaken mindfully can be a powerful form of meditation. That is why Zen retreats include working as well as meditation. No work is menial. What we do is not what we are.

". . . Baso stayed on the same mountain doing nothing but zazen day and night. One day Master Nangaku asked Baso, 'Sir, what are you doing here?' 'I am doing zazen,' answered Baso. 'What do you hope to accomplish by doing zazen?' asked Nangaku. 'I am only trying to be a Buddha,' replied Baso. Hearing this Nangaku picked up a brick and started to polish it. Baso was surprised at this and asked Nangaku, 'Why are you polishing that brick?' Nangaku replied, 'I am trying to polish this brick into a mirror.' Baso asked again, 'How can you polish a brick into a mirror?' Nangaku shot back, 'How can you sit yourself into a Buddha?'"

—SHIBAYAMA

Gandhi provides a compelling example of this. He wore only homespun cloth. He spun his own cotton yarn and taught others to do the same. Making cotton into yarn was on one level a way to break the British monopoly (the British were harvesting cotton grown in India, shipping it to Britain to be spun and woven into cloth, and shipping it back to India to be sold at a tidy profit). But it was also testament to his understanding of fundamental things, exercising majesty through simplicity and finding nobility through work. Imagine the most powerful leader of a great nation spinning yarn in his spare time. Does that really seem foolish or demeaning? Then ask yourself this: Is it scandalous? Does it need to be publicly denied? There are probably worse things leaders can do in their spare time.

Not everyone can be Gandhi, but if you're the boss you can count on having one thing in common with him: you'll pay a price for being an authority figure. The "loneliness of command" comes with the territory. To keep the weight of authority, you need a little distance from those who work under you. You can no longer fraternize in the same way. An office softball team or a drinks date after work helps create social bonds, restoring some of the humanity offices tend to strip away, but when you're the boss you have to stay a bit aloof—without stepping over the line into snobbery. At the same time, you have to avoid isolation so you won't lose touch. Even more importantly, you don't have as many others to lean on when you're running the ship. The pressures on you are the greater when you have no one to share them with. The burden of leadership is not for everyone.

ETHICS

Another common work issue is facing an ethical dilemma. You'll read about ethics and morality more comprehensively in Chapter 11, but here we'll look at a few factors specific to work. On the most basic level, to live an ethical life you need to make sure your work is ethical too. One of my clients quit her journalism job because, as she told me, she was tired of making up stories. Sheila had gone into journalism trained in the ideals of objective reporting. But on the job, her editors didn't just assign a story; they told her what the content of her story should be. Her reporting was just a way of corroborating a preconceived idea, and that flew in the face of all her ideals. She felt that the guiding force behind her organization (and we're not talking about the *National Inquirer*) was marketing, and reporting facts or discovering truth was nowhere on the radar screen.

Everyone's ethical warning lights go on at different levels, and Sheila felt she was being asked to compromise too much. Though working will always involve compromises, it is important to know when those adjustments take you over a line you don't want to cross and to take action to stay on the right side of that line.

In the corporate world, ethics has too often become the purview of the legal department. But being legal doesn't make something moral, and for that matter, being moral doesn't make it legal. Legality includes everything the law permits or doesn't expressly forbid. Morality is an even older idea, predating—and supposedly informing—legislated laws. Our laws are a reflection of our morals, which often draw on the earlier codification of morals into spiritual laws: organized religions.

But making something legal doesn't make it right. By all means dot your i's and cross your t's the way the folks in the legal department tell you to. But if something is making you morally queasy, don't let legality alone appease you. The lawyers are going to be satisfied with anything that doesn't expose the company to liability. Your personal standards should be different. Evaluate the ethics of the situation, factoring in your obligation to your employer, your personal responsibilities, your professional code of conduct, and your commitment to doing a good job. (You'll read more about how to do that in Chapter 11.) Then take the necessary actions, if any, with the great philosophers as your guides.

The loyal-agent argument ("I was only following orders") won't absolve you if you make a misstep, so you have a fine line to walk between evaluating what you think is right and fulfilling your commitment to your employer if that employer is asking you to disregard the ethical implications of a situation. It isn't easy to kill or buy off your conscience. But everyone does have a price. And once you've sold your soul—your virtue—you can't buy it back.

Chinese philosophy teaches that a hallmark of the right thing is that by doing it, you remain blameless. If you satisfy that criterion, you are in a moral realm. Another philosophical first principle is from Jain and Hindu thought, where it is called *ahimsa.* That translates into English essentially as "nonharm to sentient beings." It is the basis of every professional code of ethics. If your actions cause harm to others, they are unethical. Systems of morals get more complicated, but if you satisfy this basic requirement, you are well on your way.

ABRA

Abra had enjoyed early and rapid success in her business career before she became disillusioned with the entire system. She began to suspect that the bottom line for all businesses was profit no matter what the cost, and staying ahead of the competition by any means necessary. For many workers, that meant mindless conformity to the dictates of those higher up in the organizational chart and a complete lack of basic meaning and value in the work they did on a daily basis. The highest level managers might be able to find some personal fulfillment, but it would come primarily in the form of a giant paycheck. As far as Abra was concerned, what you had to do to get to payday couldn't be considered either healthy or humane.

While on maternity leave, Abra began to see that increasing affluence was creating a stifling trap for her. Although she was apparently succeeding at work, an expensive lifestyle and a never-ending quest for illusory status built a cage (albeit with golden bars) locking her into a job that paid for the mortgage, the car, the personal trainer, and the complete renovation of her home—though she spent less and less time there. The job paid well, but it cost her too: its price was her contentment. Abra's short recess from the rat race had given her a new perspective, and she quickly decided that she wouldn't go back after her planned six weeks after all. In fact, she wanted never to go back.

Her boss and colleagues couldn't believe she would walk away from her corner office, but to Abra her choice was suddenly clear. She had no doubt that she had done the right thing for herself and her family, but she sought philosophical counseling with my colleague Stephen Hare because she felt adrift without life goals to replace her former ambition—being president of the company she worked for. She felt there was something beyond her family responsibilities driving her, but she couldn't identify it on her own, and without knowing her goals, she didn't know how to work toward them. She felt generally disoriented in the present and perplexed about the long term.

Abra's first task was to examine her assumptions about the business world and about her need for a calling beyond parenting. She had already turned to some philosophical texts seeking answers (a throwback to her undergraduate liberal arts education, before it got mothballed in favor of the M.B.A.), so she was a good candidate for bibliotherapy—reading the philosophers herself and then discussing their ideas with her counselor. (That's not for everyone, and certainly not necessary in order to benefit from philosophical insights in your own life.)

She started with Henry David Thoreau's *Walden* and E. F. Schumacher's *Small Is Beautiful: Economics as if People Mattered*, which critically examine economic assumptions that underlie our society. These books buoyed her convictions about contemporary capitalism's soulless aspects, helped her refine her own critiques, and provided hopeful portraits of alternative arrangements. Of course, if she had turned, say, to Ayn Rand, Abra would have been reading critiques of her own critiques, but that's why finding philosophers who mesh with your perspective is important. Thoreau himself wrote of the power of books to change lives ("How many a man has dated a new era in his life from the reading of a book") and about marching to your own drummer, so before he even got into economics and social structures, he was a good match for Abra.

> *"The mass of men lead lives of quiet desperation. What is called resignation is confirmed desperation. . . . It is characteristic of wisdom not to do desperate things."*
> —HENRY DAVID THOREAU

With confirmation of her ideas about the downside of business and inspired by the utopias sketched in the books she read, Abra began to think about the way she thought society should work in an ideal state. The finer a point she could put on that, the better prepared she would be to live it. In doing so, she was calling on her own system of ethics and morals (a process detailed in Chapter 11). At the same time, Abra reconsidered whether she really needed a calling of some kind in her life, or if that was unnecessary baggage she was carrying out of the corporate world.

Recall that Aristotle recommended contemplation as the greatest happiness, which made sense to Abra. Many other philosophers have done so too. Abra decided to enroll in philosophy classes at the local university—but primarily for her personal use, not with an eye toward a degree or professional goals. She also began volunteering at organizations that were working to make the world a better place—a place more like the one Abra dreamed of living in. Though her immediate goal was primarily to be a mother, she used her volunteering experiences to decide that when she was ready to reenter the workplace full-time, she would choose a similar company—that is, one more clearly driven by something other than a pure profit motive.

Despite her reservations about the capitalist system we live with, Abra realized that it still provided opportunities to engage fully on a professional level without feeling you'd sold your soul to the devil. She also realized that the system would change only in the direction people worked to change it. The backing of other philosophers gave her confidence in her own stance to take on that challenge. The first step was setting her course by her own stars.

> *". . . to me no fortune seemed favorable unless it afforded leisure to apply oneself to philosophy, that no life was a happy one except insofar as it was lived in philosophy."*
> —AUGUSTINE

10

Midlife Without Crisis

"To every thing there is a season, and a time to every purpose under the heaven."
—ECCLESIASTES

"Time is but the stream I go a-fishing in."
—HENRY DAVID THOREAU

STELLA

As she turned fifty, Stella took stock. She saw a long career as a legal secretary, with a position of security and value to the firm. She saw a long marriage and then a lengthy solitary stretch after her husband died. She saw grown children making headway in their own lives. In general, she liked what she saw. She also decided to make some changes.

Stella realized she had always been very controlling and aggressive in her relationships and, since being widowed, had used relationships primarily for sex. She knew that most people, aside from her children, perceived her as being emotionally cold. She knew how important her few long-term, intimate friendships were to her and that she could form close ties given enough time to get to know someone. Now she felt ready to establish a more lasting and emotional relationship with a man. She knew she would be fine on her own—she had been for nearly ten years—but resolved to reach out to one of the men she occasionally dated to see if they could build something more stable and satisfying.

At the same time, Stella decided that after nearly thirty years of working for a corporation, she wanted to work for herself. She had studied art therapy as a form of creative expression and stress reduction over the years, and now she completed training to be an art therapist. She left her administrative job and started running art therapy workshops at community centers, assisted living centers, and social centers for seniors. She spent many anxious weeks anticipating this major transition but trusted in herself and her abilities. Soon her schedule was booked for months in advance. Though she was not earning as much as she had at the law firm and was working almost as many hours, she was enjoying herself immensely and felt she was helping her students and her community.

That's no midlife crisis! Serious introspection, profound change—but not a whiff of panic. The phrase *midlife crisis* generally conjures up a man with thinning hair buying a little red sports car and leaving his wife for a much younger woman while remaining unhappily in whatever well-paying but stressful or unchallenging career he's pursued all his life. Though most people do experience upheaval during their middle years—including the variety that makes you trade in what you have for a supposedly sexier model—*crisis* is a misnomer. Change is a natural part of the life cycle, but there's nothing in the definition of change that necessitates crisis.

When approached with an appropriate philosophical disposition, midlife should present an opportunity for personal growth and for fine-tuning aspects of your life that you may have neglected to your detriment. Stella's story shows that even the most earth-moving changes can be faced with equanimity. Stella actually came to consult me not only about the changes she was making in her life, since she apparently had a handle on those, but about a way to face her own mortality. (That topic is covered in Chapter 13.)

CHANGE, NOT CRISIS

So that's my pitch for replacing *midlife crisis* with *midlife change* in our shared vocabulary. Gail Sheehy brought the topic out of the closet with *Passages*, in the 1970s. Although that book takes a psychological rather than a philosophical approach, the title has the positive connotation I think is appropriate. Wherever the journey takes you, there's no need to think of it as an emergency or a calamity. You can't step in the same river twice, as we learned from Heraclitus. (One of his more observant students is said to have commented that you can't even step in it once! It changes even as you step.) But wouldn't you rather have refreshment from a flowing stream than a stagnant puddle? It would certainly be healthier, and probably tastier too. To stay clear and fresh, the water must keep moving. So must you.

In the West, we have thought comparatively little about life stages. We've been pretty much satisfied with the categories of childhood and adulthood, adulthood being a vast expanse of undifferentiated territory, with perhaps one subdivision for "senior citizens." Women have the mixed blessing of more identifiable biological life stages: menarche, pregnancy, motherhood, and menopause. The acceptance of periods of life where energy is differently focused helps avoid the kind of crisis mentality that can take over if change approaches unexpectedly. Then again, the one-dimensional tie to the reproductive cycle doesn't take into account the many facets of every person's life. And many of the stages have a strong negative connotation, as in the discomfort of menstruation, the pain of childbirth, and the undesirability of aging.

Eastern thinkers have a tradition of seeing life in phases for all people and for taking a more holistic, positive view of each stage. Confucius, for example, recognized that different periods of life focus one's energy differently.

> *"At fifteen, I set my heart upon learning. At thirty, I had planted my feet firm upon the ground. At forty, I no longer suffered from perplexities. At fifty, I knew what were the biddings of Heaven. At sixty, I heard them with docile ear. At seventy, I could follow the dictates of my own heart; for what I desired no longer overstepped the boundaries of right."*
> —CONFUCIUS

Traditional Hindu thought lays out four stages of life: the student stage, the householder stage, the stage of retirement from worldly affairs, and the stage of perfect detachment from worldly strife.

Without a familiar framework to map out the stages of life, we've been in crisis mode when facing midlife change. Acknowledging that we all go through phases allows us to put markers along the long road of adulthood without crashing and burning at each expanse of new terrain. We already have narrowly defined stages through the earliest part of our lives. One of the most popular baby care books breaks down a child's life month by month, specifying what a parent can expect almost to the week. As children grow, our society recognizes a transition every year—grade levels—and we even use labels as specific as *freshman* or *sophomore.* Our culture looks for a few other landmarks— marriage, becoming a parent—but then there's just the open road. There's no label for someone forty-two, twenty-one years into a career, and seventeen years into a relationship, with teenage children and a general sense of malaise.

There is no definition of *midlife* and certainly no specific kind of change that comes standard as we move through adulthood. The focal point could come at thirty-five or forty-one or fifty-seven—or never. For many people there will not be just one major transition. There may be several big transitions or a series of small adjustments. But with life expectancy longer than ever before in human history and people remaining healthy and active later and later in their lives, major shifts are in store for most people. Preparing for them, or addressing them, by elucidating your own philosophy is the key to making the most of whatever comes your way.

GARY

Gary's road was rockier than Stella's. After a long and successful career as a physiotherapist, Gary saw burnout on the horizon. He was tired of the stress, tired of hospitals, tired of the intensity of his work.

He knew he wanted, and needed, a change. But he came to talk to me because he didn't know what to change to. If he stopped working as a physiotherapist, he didn't know what else he would do. Having no other prospects didn't scare him, as it would many people, but it perplexed him. He couldn't shake the feeling that he *should* know what was next for him. He sought philosophical counseling not so much to figure out what to do next as to figure out why he couldn't figure out what to do next.

That may sound complicated, but for Gary the path to understanding was an easy one. This is another instance where I asked a client to reevaluate the story he was telling. I helped Gary check his assumptions. Why should he know what the next phase held? No one has perfect knowledge like that, save God or the Fates, if you believe in them. Who was Gary to play God?

As long as Gary was thinking like a strict rationalist, he was uncomfortable with his situation. Strict rationalists believe we can figure out everything using our reason. To such a rationalist, the world is cosmos (a Greek word meaning "order"), not chaos. The world makes sense. Since Gary assumed that he should be able to figure out exactly what to do next but couldn't actually do so, he thought there was something wrong with him. However, Leibniz's brand of rationalism seemed more applicable to Gary. Leibniz asserted that while every state of affairs is brought about by some "sufficient reason" (i.e., doesn't happen by accident), we can't always know exactly what that reason is. Humans can't always figure everything out. The cosmos is just too complex for us.

". . . no fact can be true or existing and no statement truthful without a sufficient reason for its being so and not different; albeit these reasons most frequently must remain unknown to us."
—GOTTFRIED LEIBNIZ

I don't think a strictly rationalist view can get you through the night any more. Humans are great at coming up with questions that can't be answered, even as we discover more and more about the universe and how it works. I asked Gary to consider the limitations of that way of thinking, starting with asking, "Is it rational to stay indefinitely in a job I hate?" Probably not. But that doesn't mean there's an immediate answer to the next question, "Then what am I supposed to do instead?"—at least not one accessible to intellect alone. Perhaps what you are supposed to be doing just now is not knowing what you are supposed to be doing just now.

All this appealed to Gary, and he paraphrased the famous passage from Ecclesiastes, "a time to every purpose under heaven," to me: there's a time to know what you're doing and a time not to know. I reminded Gary of another of passage from Ecclesiastes, repeated at every opportunity by Ecclesiastes himself: "All is vanity, and a striving after wind." Thinking you know exactly what you're doing is one kind of vanity; knowing that you don't know, but thinking you should know, is another kind.

Gary felt more comfortable with a revised rationalism: things have an explanation, even if we don't see it at the time. If you are experiencing change or uncertainty or unhappiness and don't know why, your task becomes to discover your purpose, avoiding the egotism of presuming you already know, or should know.

"To every thing there is a season, and a time to every purpose under the heaven: a time to be born, and a time to die; a time to plant, and a time to pluck up that which is planted; a time to kill, and a time to heal; a time to break down, and a time to build up; a time to weep, and a time to laugh; a time to mourn, and a time to dance; a time to cast away stones, and a time to gather stones together; a time to embrace, and a time to refrain from embracing; a time to get, and a time to lose; a time to keep, and a time to cast away; a time to rend, and a time to sew; a time to keep silence, and a time to speak; a time to love, and a time to hate; a time of war, and a time of peace."

—ECCLESIASTES

There was even a time for The Byrds to make a great hit record ("Turn, Turn, Turn") out of these lyrics. From that perspective, Gary began to see that having a period of time without—for once—a clear direction could be of great value. Up to this point, one thing had led directly to another for most of Gary's life—college, graduate school, first job, specialty training, promotions, and so on.

I told Gary about Norman Spinrad's science-fiction story "The Weed of Time," in which a man eats an herb that gives him the power to see both everything that has happened and everything that is going to happen. His life becomes a torture. If your college literature professor had asked for an essay on this story, your topic sentence (if you wanted that A) would have looked something like this: Discovery is part of the joy of living.

Gary found turning down a road without a clear map both anxiety-producing and exhilarating. His footing was less sure, but he had a wonderfully expanded vista. Gary chose to see his uncertainty as a great gift and was determined to use it well. He eventually left his hospital job without having a specific plan. What he did have was the blessing of his life partner, Mary, who had a good job she was happy with. Gary's happiness was important to Mary as well as to Gary. She didn't want him to take a job just for the sake of having a job any more than he did. Mary earned enough to cover their expenses, so Gary had the opportunity to find his right path relatively free of financial pressures.

Looking at his situation through the PEACE process, Gary had already worked through the first three stages by the time he sought philosophical counseling. He had clearly identified his problem: needing to change his career. He was experiencing the associated emotions—anxiety first among several—without being particularly debilitated by them. He had analyzed his apparent options but was unable to answer two fundamental questions: "What am I supposed to do next?" and "Why can't I figure out what I'm supposed to do next?"

His breakthrough came at the contemplative stage, looking at his undefined future as an opportunity rather than a barrier. His philosophical disposition changed from "There must be something wrong with me because I can't find answers" to "I'm fortunate to be entering a phase of my life in which I don't need answers." Given time for reflection, he'd be able to refine his own philosophy and contemplate options for work consistent with it. But he'd do so with equanimity instead of anxiety. Leibniz, not Librium!

Starting with the idea that wisdom is not the same as knowledge and that not knowing what we're doing doesn't mean we're not doing the right thing, Gary was on the path to finding answers by not looking for them. Sometimes your purposes may be best served by not knowing. If we could explain the cosmos by reason alone, there would be no need for scientific experiments. But when our pure reason fails, we need to conduct experiments, without which some of the greatest discoveries of all time—from electricity to polio vaccine—would never have come to light. Sometimes answers aren't immediately forthcoming in your life either, which means that you might benefit from experimenting too.

GAUGUIN AND YOU

For some people, like Gary, the crisis comes when they don't know what is coming next. For other people, knowing what is coming next can provoke the crisis. For those who complain that the world is too predictable, having life scheduled feels inauthentic and unbearable. If you're one of them, you'd trade security and prosperity for a path of risk-taking and discovery. You don't want to live your life according to a long list of preset rules. If that's you, you'll have to keep one eye open for the "Gauguin problem" as you chart your course.

Paul Gauguin abandoned his successful banking career, loving wife, and growing children when he fled to Paris to study painting, then moved to Tahiti to pursue his calling. Those of us who love Gauguin's works are grateful for the great gift of art he left to the world. But did Gauguin behave morally in regard to his family and his business? When, if ever, do ethics and obligations become subservient to greater creative goals? You'll have to draw that line for yourself, but you should be aware of where it lies as you balance stability and exploration.

One of the people I admire most is a sort of reverse Gauguin. After a career as a philosopher and logician, Stephan Mengleberg served as assistant conductor to Leonard Bernstein at the New York Philharmonic for some years. Aside from being Leonard Bernstein, this was surely one of the sweetest positions in classical music. But in his fifties, he heard that little voice murmuring inside—and it was calling him to the bar. Not that he took up drinking: he became a lawyer and practiced happily and successfully for the rest of his life. He continued tutoring music students too but had the courage to be true to himself, independent of his age, even when that meant a drastic change of course. Stephan was, to my knowledge, the oldest person ever to join the bar in New York State, at age fifty-nine.

I knew a woman named Judy who, long retired from the civil service, went back to school and earned a B.A. in history—a subject she had always enjoyed but never pursued—graduating at age eighty-five! There's no reason to let arbitrary barriers of age or stage of life stop you from doing whatever is in your heart.

ANN

Ann faced a Gauguin-like choice of a more free-spirited, creative life—but feared the loss of security it might bring. Like Gary, she excelled at a job she could no longer stand. She was bored to tears as an administrative assistant at a health center and had recurring conflicts with one of her bosses. Now a new, tempting job offer had fallen into her lap, yet she was undecided about taking it. So she sought out philosophical counseling to help her decide whether she should accept the new job—and understand what was holding her back. Though she was younger than most people are when they face midlife change, the overarching questions related to her immediate dilemma put her in good company with all those a few years ahead of her who were wrestling with similar issues. Her conflict stemmed from her career but had implications for all areas of her life.

During the past year, Ann had spent one evening a week tutoring in an after-school program in her neighborhood. She had formed a close bond with one of her students, and the girl's parents wanted to hire Ann to teach their daughter at home full-time. This would mean working six hours a day during the school year only, at the same salary she was earning at the counseling center. In addition to the prospect of a lot more free time than she ever had in the 9-to-5-plus-overtime grind and without the commute through bumper-to-bumper traffic, Ann was drawn to teaching as a way to use all her skills and help someone else at the same time.

But Ann was fearful. The student's parents had offered a one-year commitment, and Ann worried about what would happen at the end of that time. Would she be rehired? Would the girl still need her? Would the parents be pleased with her progress? Even in the best of circumstances, the girl would be headed for college in a few years. Ann enjoyed a happy, supportive relationship with the man she lived with but didn't want to be financially dependent on him. Raised with a strict work ethic, she felt vaguely that sinfulness was associated with not being busy working all the time. She felt secure, if unfulfilled, in the job she had and wasn't sure she should risk that for this new venture.

Humans, like most organisms simply trying to survive, have a fear of the unknown. It can be a useful fear: if you don't recognize something, you don't know if it's safe, so you'll be better off where you know you are safe. Of course, you also don't know if whatever it is might be more safe. And what if you aren't safe where you are? The survival instinct only gets you so far. Being ruled by such a conservative approach works to keep you alive, but it offers no guarantees about satisfaction or fulfillment from the life you preserve.

Perhaps that's why humans also seem to desire the unknown. The promise of discovery makes us feel alive. We seek security, but as soon as we have it, we risk it. Watch a toddler look around for his mother as soon as he ventures a few steps away—then continue further afield once he sees she's still there—and you'll see what I mean. We live with a constant tension between the comfort of security and the excitement of new experiences.

Ann was by nature conservative, but even she felt the pull of the expanding horizon. She felt she was missing a challenge in her life and that the corporate routine was choking her ability to be fully herself. Less individual expression is the price of the security that institutions provide. Ann was beginning to think the price had become too dear but was struggling to establish the value of both the security and the self-expression. One side of the balance held opportunity and freedom, the other predictability and permanence. She wanted me to help her weigh them.

Hindu and Buddhist philosophy maintain that permanence and security are illusions. Such illusions breed attraction in the grasping mind; attraction foments desires; desires give rise to attachments; and attachments to suffering. The Stoics, as we have seen, would agree that pleasure can turn into pain quickly if the pleasure comes from attachment. We do get attached to things, even to negative things, just as Ann was attached to her stifling job. We choose the devil we know over the devil we don't know. It's a comfort that slowly poisons us, but we become so habituated to it we don't even taste the poison. After long periods of knowing nothing else, prisoners come to fear the world outside their cells. If all the gates swung open one day, many would stay right where they are.

> *"Those external relations which bring cold and heat, pain and happiness, they come and go; they are not permanent. Endure them bravely. . . ."*
> —BHAGAVAD GITA

> *"Those who are fearful when there is no cause for fear, and feel no fear when they should, such people, embracing erroneous views, enter the woeful path."*
> —BUDDHA

Her door was open, but Ann was hesitating at the threshold. The Hindu and Buddhist teachings made sense to her, and she felt they applied with even more urgency in our fast-paced world. Things evolve so rapidly now that change itself is viewed as a virtue. We're called on to face our innate fears courageously and learn to embrace change. If Aristotle ran GM, as a recent book title would have it, he'd view positive change as the Golden Mean. He would encourage Ann to accept the new opportunity since that was the middle way between the extremes of slavishness to her present job and just walking away with no plan for the future.

Philosophically, Ann was walking the fine line between free will and determinism. She hadn't taken any concrete steps to find a new job or any other way out of her current unhappy career. But she had put herself in the path of the opportunity she now faced. Volunteering was one way she had of being true to herself and following her inner calling. She had seemed to be following, by default, a fatalistic outlook—not taking action to improve her situation, as if any action would be futile. But she found good results from acting on free will (volunteering), even though she didn't foresee or intend those results. Resolving this internal debate over how much effect she herself had on the events of her life would help Ann move forward accordingly.

As we saw, Machiavelli managed the free will versus determinism clash by declaring the two ideas equal partners in getting things done in life. He advised, then, not wasting time trying to change what you can't (what is predetermined) but working on the portion you can influence. If he hadn't been so concerned with establishing his reputation as a prince of a bad guy, Machiavelli might have seconded the famous Serenity Prayer: "God grant me the serenity to accept the things I cannot change, courage to change the things I can, and wisdom to know the difference."

We humans are manipulative creatures. But we can't manipulate the whole world, and we don't need to. Artists often have the kind of faith in themselves that all of us could benefit from, the courage to navigate by their own stars. The answer to "What should I be doing?" usually comes from looking inward, not outward. Answering the call, whatever it may be, will eventually bring opportunity, as it did for Ann. Opportunity doesn't knock only once: it knocks constantly. But we are often deaf to it, or pretend not to hear it, or are afraid to answer the door.

Ann took comfort in the notion that success and failure are both, as Kipling wrote, "impostors." With the worst-case scenario of being able to learn something she could use later in life even if the teaching job didn't work out over the long term, Ann decided to quit her job and become a full-time tutor. Though she was very nervous approaching the change, after a couple of months she loved her new position and her new life.

Make the best of change, and you'll get the best from life. Forget about midlife crisis. You can experience a crisis every five minutes, or you can live five lifetimes before you even reach your middle years. The choice is yours.

11

Why Be Moral or Ethical?

A New York City police officer recently made the evening news with
the following astonishing act: upon stumbling across $35,000 in laun-
dered drug money while alone on his beat, he quickly scooped it up
and . . . turned it in as evidence! The media jumped all over this "Man
Bites Dog" story. Reporters were tripping all over themselves to praise
his honesty. The mayor presented him with a medal for integrity.

I was cheered by the story too—having heard more than enough
about cops pocketing cash—until I heard the officer talk about why he
did what he did. He'd thought about taking the money, he confessed,
but then realized that his pension was worth a lot more. He said he
didn't want to risk losing his pension if he was caught. "Why jeopardize my financial security for thirty-five grand?" he reasoned. This gave
me pause. I wondered what this cop would have done if he'd found a
stash worth more than his pension. If he followed his own reasoning,
he would have grabbed it up without a second thought.

If the mayor wanted to hand out medals, he should have inscribed
candor rather than *integrity* on the one for this guy. The officer did at
least have the courage to speak the truth. But his moral reasoning isn't

what I'd hold up for my children. What he was really saying was "I'll obey the law as long as I get more out of obeying it than I would by disobeying."

For me, this wasn't even the truly scary part of the story. What alarmed me was that no one else seemed to detect the flaw in what the man in blue had said. I seemed to be the only one who noticed, the only one who was concerned that doing the right thing for the wrong reason doesn't make you a person of integrity. Your motives as well as your deeds must be honest. Integrity involves unstinting loyalty and duty to principle, not cold calculation and expediency. Taking money that belongs elsewhere is wrong, independent of the amount involved. This story was about a potentially dishonest cop whose price just hadn't been met yet.

I don't want to discredit him entirely since he did turn the money in. In a more generous mood, I'd say perhaps he just bungled his explanation to the press. But you should be appalled by the low moral standards we have come to embrace. The rule involved here—don't steal—is a basic one that should be better understood by society in general, including law enforcement officers and the media. Yet the media were more than ready to lionize this person without ever considering the essence of what he'd done and said.

The aim of this chapter is to help you understand and apply your own system of ethics. When you get your fifteen minutes of fame, I want you to be able to give a compelling answer to the question "What was going through your mind when you decided to do that?" In many ways, this is the key to most of the situations described in Part II of this book. No matter what your specific issue, your struggle is to identify and occupy the moral high ground, do the right thing, and be able to explain to any and all comers (including yourself) why you made the choice you did. As we've seen, there are particular philosophical angles to consider as you contemplate beginning or ending a relationship, making a career change, or managing a complicated family life. But when push comes to shove, the underlying question remains: How can I act in this situation in accordance with my effort to lead a good life?

This chapter helps you get there. I've included some case studies, as usual, to give you an idea of particular ethical or moral problems and how to resolve them. But all that is presented in this chapter could be brought to bear on many of the topics in this book.

MORALS AND ETHICS

Everyone throws around the labels *moral* and *ethical,* often using them redundantly as if to make a stronger case (e.g., "He behaved both morally and ethically"). If you ask people the difference, most won't have a clue; they've just been repeating the formula because it sounds good. But we can make a distinction between the two that I think is useful. Ethics refers to a theory or system that describes what is good and, by extension, what is evil. Mythology and theology are the oldest sources of ethics, though philosophical systems are often more discussed today. Morals refers to the rules that tell us what to do or not to do. Morality divides actions into right and wrong.

Morals have to do with your personal life: What is appropriate behavior on a first date? Is taking a ream of paper from your corporate office home to your kids a crime? Ethics are more theoretically focused: How do we judge white-collar crime versus violent crime? How do we allocate transplant organs as long as demand outstrips supply? Morals are the rules you live by; ethics are the systems that generate those rules.

Ethics are about theory, while morals are about practice. Your personal philosophical standing will be strongest when you successfully link the two. If you know what is good or evil, you should be able to figure out what is right or wrong. You have to know your options, weigh the pros and cons, and find a way of reasoning morally about what confronts you so that you feel justifiably right in your intended response. If you don't feel right, maybe you shouldn't do whatever it is you're contemplating doing. If it is the right thing, there will always be a way to justify it. Note that rationalization is something else altogether. You can rationalize anything, shaping and misquoting any idea to fit your plan (no one will find out; nobody's perfect; the devil made me do it; God will forgive me; I'm the president). *Justification*, however, shares the same root as *justice* and *just*. It demands weightier deliberation and in return provides surer footing.

Your challenge is to have a personal ethical system up and running that you can rely on for moral guidelines. You'll have to begin by contemplating what is good and what is bad. That problem has stumped philosophers down through the ages, so don't expect to have a complete and infallible answer by the end of this chapter. In *The Republic*, Plato posits a dialogue in which Socrates asks him to define the Good: "Is it knowledge, or pleasure, or something else?" He'd already pinned down several virtues, including temperance and justice, but faced with this challenge, Socrates replies, "I am afraid it is beyond my powers."

Centuries later, the view hadn't gotten much clearer. "'Good,' then, . . . is incapable of any definition, in the most important sense of that word," wrote G. E. Moore. Nietzsche complained of the "ancient illusion called Good and Evil." Like others who have tried before you, you may not be able to answer this riddle precisely. Nonetheless, it is crucial to get your feet wet trying. It is the only way to create a strong foundation.

Plato held that people have an intuitive grasp of the Good, though we have only poor copies of the ideal in our real world. "The highest object of knowledge is the essential nature of the Good, from which everything that is good and right derives its value for us," he wrote. As we've seen, however, Plato didn't reach his own highest goal and never pinned down a definition.

Hobbes took a different view: "Whatsoever is the object of any man's appetite or desire; that is it, which he for his part calleth 'Good': And the object of his hate, and aversion, 'Evil.'" In other words, Hobbes opposes Plato and says that there is no universal essence of good; *good* and *evil* are only labels that we use to describe our likes and dislikes.

The Tao teaches that we can only recognize good in comparison to evil but doesn't provide any working definition.

> "... the highest object of knowledge is the essential nature of the Good, from which everything that is good and right derives its value for us."
> —PLATO

> "... these words of Good, Evil ... are ever used with relation to the person that useth them: There being nothing simply and absolutely so."
> —THOMAS HOBBES

Why be good, or moral, or ethical? Why be concerned about right and wrong? What's in it for us? All of this is easier if you subscribe to a religion that lays out good and evil for you, on the authority of God. The great religions all give moral guidance originating from a divine power. Attributing the rules to God kills two birds with one stone: you've got very specific morals to guide your actions and an absolute ethical system to hang them on. Obeying God's commandments is what it means to be right. The rules come from God, and God is good.

If that works for you, you are ahead of the game. Even if you don't believe in a religion, you can still use the wisdom the ancient theologians presented, without owing allegiance to a deity. The scriptures of all major religions contain profound insights into morality from which anyone can benefit. But to gain philosophical resolutions to life issues, with or without faith, you'll have to seek out and understand the important precepts and work them into your personal worldview. You've surely heard before that you should "walk the walk" as well as "talk the talk." While that is good advice, I'm an advocate for talking the talk as well as walking the walk. The thought and reasoning behind your actions is a key to facing and resolving whatever comes your way.

SCIENCE

Religion is just one way to find the ethical and moral rules of the road. Many people prefer the substitute god Science. A scientist like

E. O. Wilson, calling for "ethics to be removed temporarily from the hands of philosophers, and biologicized," can drum up a substantial hallelujah chorus, especially on a university campus. The idea is to look for evolutionary explanations for our behavior—to show how good behavior is supposedly favored by natural selection while bad behavior is supposedly disfavored. We'll be able to distinguish Quakers from Nazis by looking at their DNA.

While I think we can learn a lot from evolutionary theory, I don't believe we'll find anything in our genes alone that compels us to be good or evil, or rather, to do right or wrong, thereby illuminating what is good or evil. Norms like "Don't marry your cousin" can be tied to biology—mixing closely related genes dramatically raises the rate of genetic abnormalities in offspring. But while sociobiology tells us that we increase our chances of passing on our genes through certain altruistic actions, we can also pass them on simply by being promiscuous. Herod supposedly had seven hundred wives and God knows how many children. While this would give him a high grade for fitness on the sociobiology index—right off the scale, I should think—I'll bet it wouldn't make him a pillar of E. O. Wilson's (or anybody else's) community.

There is a difference between what is good for us and what is good in a universal or ideal sense. I don't believe that science will ever take us there. In fact, the natural origin of the incest taboo is the only consistently workable example of morality derived from science. (And that still begs the question of why so much incest takes place, despite the taboo.) We clearly have—and need—many more morals to shape us and our societies. The origins of morality are far from clear, and far from evolutionary theory.

WHAT IS GOOD?

Science and religion both contain nuggets of moral truth, regardless of whether you sign up for their complete programs. But if they don't satisfy you on their own, another means of approaching morality and ethics is secular philosophy. "What is Good?" is perhaps the oldest question in philosophy. Western philosophy offers at least three main ways of thinking about the answer: naturalism, antinaturalism, and virtue ethics. Each comes in several varieties.

The foremost naturalist is Plato. He founded the essentialist tradition, which holds that there is a universal Form that is Goodness. For Plato, a Form is an idea, not a material thing, but real nonetheless. He divides the world of appearances—concrete things as we observe them—from the world of ideas, or Forms. All things on earth are copies of Forms, and while the Forms themselves are perfect (that is, ideal), the copies are necessarily flawed. There is, according to Plato and his followers, an ideal of Goodness. To be moral beings, our job is to copy the ideal as best we can. As time passes and we gain understanding, we should be able to make better and better copies, drawing closer to the ideal of Goodness all the time. In the realm of ideas, Goodness plays the role of the sun: its radiance illuminates all the other Ideas too.

Plato, however, does not—and says he cannot—provide a specific definition of Goodness. He believes that the mind can apprehend the essence, even though it can't be put into words. This gets circular—a good person is a person full of this (undefinable) essence; a person full of this (undefinable) essence is a good person—so to get on board you have to make your peace with intuitive, rather than explicit, knowledge of Goodness.

Plato firmly believed that ethical education was crucial to moral behavior. He insisted that critical thinking skills (in his day, that meant Euclidean geometry) were necessary precursors to moral reasoning. He'd be appalled, then, at the way we teach ethics to very young children—if we do so at all. If Plato could assess contemporary American education as a whole, he'd find it ethically impoverished and morally bankrupt.

We might do well to follow Plato's advice and lay down a foundation in critical thinking and mathematics before we dive into ethics. Minimally, we should teach how to reason about cause and effect. If you are a parent of a young child, just consider how many times a day you hear yourself saying, "We don't do X," or "Be a good girl and . . . ," or "That's bad!" True, you can't give a two-year-old a discourse on all the reasons why. But as your children grow older, you need to provide more of the whys and wherefores and help them develop their own skills in reasoning morally about their actions, or your rules will just be a list of seemingly random regulations. The schools won't do this for you anymore, and without it, your children won't have the ability to manage themselves morally, which personal and social maturity requires. You won't get compliance either!

Insofar as human sociobiologists like Wilson believe that ethics arise from nature, they are also naturalists, although they need not share Plato's essentialistic views. And religions, too, are naturalistic, for they attribute Goodness to God, who presumably confers it on us.

A second major Western philosophical school of thought about "What is Good?" is antinaturalism, which also comes in many varieties. Antinaturalism generally states that nothing in nature is either good or evil. That is, what is natural and what is moral are distinct. Hobbes, a nominalist, is an important proponent of this school. As we have seen, nominalists maintain that there are no universals, that good and evil are just names we give to things. There is no good and bad, he'd tell you, just people's likes and dislikes. Morality, in practice, is narrow, personal, and subjective. No two people will completely agree on the ground rules, which explains why we come into conflict so easily.

G. E. Moore, another important antinaturalist, believed that while there are many things we can measure with instruments, Good isn't one of them. Rather, Good is undefinable, unanalyzable. When we try to assess it, we commit the "naturalistic fallacy." Moore doesn't recognize any detectable essence of Goodness. No one can say what Good means, he holds, and it certainly isn't just a matter of labeling things (to differentiate his point from Hobbes's). Moore believed that there are right and wrong actions but that they cannot be derived from any concrete conception of Good.

> *"'Good,' then, if we mean by it that quality which we assert to belong to a thing, when we say that the thing is good, is incapable of any definition, in the most important sense of that word."*
> —G. E. MOORE

Hume anticipated Moore's line of thinking. He held that you can never "derive ought from is"—meaning that you can't draw any logical conclusions about what should be done simply from what has been done. For instance, just because X harms Y, it doesn't follow that X was wrong to harm Y. That would follow only from the additional assumption that harming another is wrong. But then you have assumed a moral principle, not proved it. Hume emphasized that although we make value judgments, we must acknowledge that they are not drawn from hard facts.

A third way of thinking about Good is Aristotle's so-called virtue ethics, which we have already seen in several cases thus far. Virtue ethicists believe that goodness is a product of virtues. If we inculcate virtues in people, they will be good. This view was also developed by Confucians and by many religious moralists.

> *"Thus it is possible to go too far, or not far enough in fear, pride, desire, anger, pity, and pleasure and pain generally, and the excess and the deficiency are alike wrong; but to feel these emotions at the right times, for the right objects, toward the right persons, for the right motives, and in the right manner, is the mean or best good, which signifies virtue."*
> —ARISTOTLE

Given the inherent limits in all of the approaches outlined above, you can see that you'll have your work cut out for you. But before you get down to it, I want to include two final perspectives from Eastern philosophy. You've come this far in an attempt to refine your own way of thinking about what goodness means, and for your trouble you've seen a lot of theory in a short time. Here's one you can put into practice immediately: the doctrine of *ahimsa*, or nonharm. A central tenet of Hindu philosophy, borrowed from Jainism, practicing *ahimsa* means acting to assure that you cause no harm to sentient beings. This is a very simple measure of good. How good you are is inversely proportional to how much harm you do to sentient beings. What harms others is bad; what is bad harms others. What helps others is good; what is good helps others.

If you've been paying attention, you'll have noticed that *ahimsa* is concerned not just with other people but with all sentient beings. The Judeo-Christian orientation doesn't generally extend to animals—after all, in Genesis, humans are explicitly given dominion over them, and we lost no time in exercising that power. I think you can usefully apply the principle by extending it to humans in the first place, and you're surely safe in assuming that the more sentient the life form, the more harm you can do to it. It is food for thought, and a point not to be overlooked as you fashion your own system. You need to know where you draw your own boundaries—and why.

Hindus recognize the connectedness of all things, which is why they do not limit the application of *ahimsa* to fellow humans. In fact, they hold that *avidya*—blind ignorance, or doing harm without knowing it—does not spare one from the consequences of doing harm (which we'll get to in a minute). Realizing your potential to harm is a key insight, and understanding how not to do harm should be your personal quest. That means taking care of what you think, say, and do.

Ahimsa is such a powerful idea that it echoes across time and around the globe. You hear it again in Hippocrates' advice to doctors, "Make a habit of two things—to help, or at least, to do no harm." And again, implicitly, in the Golden Rule: "Do unto others as you would have others do unto you." It appears in Matthew 7:12: "Therefore all things whatsoever ye would that men should do to you, do ye even so to them: for this is the law and the prophets." Similarly, Hillel wrote, "What is hateful to you do not do to your neighbor. That is the whole Torah. The rest is commentary." And Aristotle: "We should behave to our friends as we would wish our friends to behave to us." And

Confucius: "What you do not want done to yourself, do not do to others." If Western philosophy can be summed up as a footnote to Plato, perhaps all the complications of ethics are footnotes to these formulations, and Hillel is right: the rest is commentary.

Hindu and Buddhist traditions spell out the consequences of doing harm with *karma*, a moral law of cause and effect. Literally, karma means "the ripening fruits of action." As the popular saying has it, "What goes around, comes around." Or as Paul expressed it (in Galatians 6:7): "for whatsoever a man soweth, that shall he also reap." Do good, and good will come back to you; do evil, and evil will return to you. The mystery is in what form the effect will come and how long it will take. We tend to forget the things we did to start the ball rolling (for better or worse), but that doesn't mean there isn't a connection. We might learn from finding the connections where to place our feet as we proceed. But even if we don't divine the pattern, believing that there is one is a powerful motivation for doing the right thing. The take-home message is that what we think, say, and do has consequences. Too often in contemporary American society, we act as if we were unfamiliar with that kind of responsibility.

> *"If a man commits a meritorious deed, let him perform it again and again; let him develop a longing for doing good; happiness is the outcome of the accumulation of merit. Even the wrongdoer finds some happiness as long as the [fruit of] his misdeed does not mature; but when it does mature, then he sees its evil results. Even the doer of good deeds knows evil [days] so long as his merit has not matured; but when his merit has fully matured, then he sees the happy results of his meritorious deeds."*
> —BUDDHA

If all this nonharm has got you wondering if you're made of the right stuff, note that when you take karma into account, *ahimsa* takes on a quality of self-preservation—hence the Dalai Lama's advice to be "wisely selfish."

Chinese philosophy takes a more practical, virtue-ethical approach to defining Good. Confucius does so most rigidly. His primary concerns are tradition, structure, duty, family, government, and maintaining society. To him, the Good is simply whatever upholds and defends those values.

In the doctrine of opposites, the Tao teaches that pure good does not exist. Lao Tzu believes that you can recognize good only in comparison to evil. Kant put this idea another way: If there were only one hand in the universe, how would you know if it was a left hand or a right hand? The celebrated swirls of the yin-yang symbol represent this by containing a small circle of the opposite color within both the black and white sides. That's a reminder that good is not the opposite but the complement of evil, and that everything contains something of its complement. In good times, be sure you side with the good and avoid evil. In bad times, your job is to find your way into the light amid the darkness.

"When all in the world understand beauty to be beautiful,
then ugliness exists. When all understand goodness to be good,
then evil exists."
—LAO TZU

By now you've probably noticed that we still haven't answered the question "What is Good?" That's because, as you can see, there is no one answer. And depending on whom you listen to, maybe the question isn't answerable at all, at least not explicitly. Unless you're ready to sign on wholeheartedly with one existing set of guidelines, there is no universally defensible ethical system that you can use to derive consistently workable morals. There are no ubiquitously conclusive arguments in support of any one ethical theory to the exclusion of all others. Notions of good are formulated differently by different people. Yet this doesn't make us moral relativists: in spite of the diversity of ethical systems, most people still believe that murder, rape, and theft (among other things) are wrong.

WHAT IS RIGHT?

So without knowing conclusively what's good, how can we know what's right? Not too easily. Even if you knew what was good, you'd still be faced with a dilemma: a choice between two main ways of understanding what is right. These two ways are called deontology and teleology.

Deontologists believe that the rightness or wrongness of an act has nothing to do with the goodness or badness of its outcome: acts are right or wrong in and of themselves. So, for example, if you subscribe to the Ten Commandments, you have a set of rules that tell you what is right or wrong. Rule books are helpful in that you can look up the rightness or wrongness of an act in advance. But rule books are also unhelpful in that almost all rules have exceptions. While most people can agree on basic rules (e.g., "Thou shalt not kill"), most people also want some exceptions (e.g., except in self-defense, or war, or abortion, or euthanasia). Thus deontologists sometimes end up killing one another over disagreements about exceptions to the rule "Thou shalt not kill." The strength of deontology is having moral rules; its weakness lies in the difficulty of establishing workable exceptions.

Teleologists believe that the rightness or wrongness of an act depends partly or even completely on the goodness or badness of its outcome. If, for example, you subscribe to utilitarianism ("the greatest good for the greatest number"), you are a teleologist. While a deontologist might condemn Robin Hood (because stealing is wrong), a teleologist would wait to see what he did with the loot. If Robin Hood opened a Swiss bank account, the teleologist would say he was wrong because he stole for his own gain; if Robin Hood gave to the poor, the teleologist would say he was right because he was helping others. But teleology can backfire too. Suppose a crime has been committed, and ninety-nine neighbors decide to round up the first stranger they see, convict him on the spot, and lynch him. They do so, and all ninety-nine sleep well that night. The greatest happiness for the greatest number can result in the greatest unhappiness for the smallest number. The strength of teleology is its open-mindedness; its weakness is its potential disregard for individual rights and due process.

So where does that leave you, on your quest to live right? With meta-ethical relativism, if you want the technical term. As we saw in Chapter 8, relativism is the belief that there is no absolute right—that some actions are more appropriate in some circumstances than in others. Meta-ethics is the comparison of competing systems of ethics: some ethical systems are more appropriate in some circumstances than in others. If you can imagine that deontology works better than teleology sometimes, and vice-versa at other times, then you're a meta-ethical relativist. Just pray that no one asks you to define *better*.

If you are facing a particular conflict right now, use it as a test case in developing a personal system of ethics. Just make sure that whatever specific result you get also works in a general sense. If you don't have an immediate moral or ethical problem, you'll still do well to prepare your theory so that it's ready when you need it in practice. All this might be something you can do on your own or with a friend or partner. Professional counseling is always an option if you need personal expert guidance or if you get stuck.

No matter how you do the work, the key to a lasting, useful way of thinking about ethics is consistency. You need to work out a system you can live in harmony with and rules you can explain to yourself and others.

ETHICS IN JEOPARDY

Your total belief system is made up of many sets of beliefs in a variety of categories: religion, politics, aesthetics, parents' beliefs, peers' beliefs, and so on. Each set is a collection of premises you hold to be true and arguments you hold to be sound, though experience or reason may modify any of those elements at any time. These sets often coexist uneasily, and conscious conflicts arise when a premise held to be true in one set of beliefs contradicts a premise held to be true in another set of beliefs. For example, you might encounter the idea that "divorce is sinful." However, you say to yourself, "My parents are divorced—and they are good people." You wonder if both premises can be true, but you're not sure you can invalidate your religious upbringing or the received wisdom of your family. You feel conflicted. You have what psychologists refer to as "cognitive dissonance." Philosophical counselors call it "existential dissonance." Regular people might say "mixed emotions."

If you have never worked out to your satisfaction a personal system of ethics, this is a sign that you need to do so. If you have done so already, don't let an apparent conflict panic you or force you to lose faith in your system. Let it prod you to explore a bit deeper, making refinements. You may want to drop one of a pair of beliefs in conflict, but that isn't necessarily called for. You can judge a problematic situation on a case-by-case basis. There may also be ways to harmonize the dissonance through action rather than (or in addition to) reason. Talk to your pastor. Ask your parents their own opinion of their breakup. Choose your own life partner carefully and put in the necessary work to maintain the relationship.

"Nobody holds a belief in total independence of all other beliefs. Beliefs always occur in sets or groups. They take their place always in belief systems, never in isolation."
—THOMAS GREEN

TED

To give you an idea of how you can put your principles into action, here's a case about Ted—a high school principal. Though the dilemma you'll read about here is very specific, I'm starting with this one because the issues and solutions are reasonably clear-cut—a luxury compared with most real-life situations. Because they can be isolated in this case, the process is more easily accessible.

At Ted's high school, students had organized a fund-raiser for a local charity. As an incentive for student participants, Ted set up a lottery with several prizes. Students received one lottery ticket for every $10 in pledges they collected. In the reception area of Ted's office, the students set up a series of boxes, each one labeled with one of the lottery prizes. Students distributed their tickets among the boxes according to their preference for each prize.

At the end of the campaign, the students gathered in the auditorium for the highly anticipated prize drawing. Ted did the honors, selecting a ticket from each box. One by one, students came up to claim their prizes—CDs, free passes to a movie theater, a gift certificate for a popular local clothing store, and the grand prize, a mountain bike—accompanied by the cheers of their classmates. At the peak of excitement, Ted announced the impressive grand total of the school's charitable contribution. He felt the assembly was a high point both for school spirit and for the spirit of volunteerism.

The next day, however, a student came to see him to report that the winner of the bicycle had not solicited any pledges herself. Rather, Tiwana had been given the ticket by Clarabel, who had raised a significant amount of money, as a token of friendship. The student's complaint brought an ethical dilemma to Ted's attention. Should Tiwana be entitled to the bike just because she held the winning ticket? Was her right to the prize compromised because she hadn't participated in the fund-raising? Word spread around the school quickly, and soon parents were calling to inquire about the situation.

But Ted was stymied. He couldn't see a just way out of the turmoil without spoiling the good feelings originally generated by the fund-raiser and prize giveaway—which were quickly wearing away under the stress of this revelation anyway. He felt so anxious he literally started losing sleep over the whole matter, so he called me for help in finding a solution he could live with personally as well as defend to others. He outlined the options he saw: Tiwana keeps the bike; Clarabel gets the bike; he conducts a new drawing for the bike, with Tiwana's ticket returned to Clarabel; Tiwana and Clarabel work out who owns the bike between themselves and a second bike is purchased and awarded to the next ticket drawn out of the box.

Ted did not want to do a new drawing (the lottery itself had been fair) or buy a new bike (thereby reducing the money going to the charity). He told me he had spoken to both girls and their parents, and all but one of them was open to the various ways of resolving the situation. Clarabel's father, however, was adamant that the bike should be Clarabel's. Even so, Ted felt intuitively unhappy with all the above options. He wanted to make the right decision but sought some ethics counseling to help him arrive at it. Together we resolved the problem.

Like many clients, Ted had guided himself through the first three steps of the PEACE process (identifying the problem, expressing his emotions, and analyzing his options) but then became stuck. He needed a philosophical hint for contemplating the best path overall and understanding the reasons behind it.

I agreed with Ted's assessment of the undesirability of redoing the lottery, so we focused on whether Tiwana should keep the bike or Clarabel should receive it. We talked over the philosophical ideas relevant to each possibility. The key insight was separating the moral claims from the legal ones. Legally speaking, the ticket (and so the bike) was Tiwana's since Clarabel had given it to her freely. Possession, as they say, is nine points of the law. (Of course, if possession had been ill-gotten—if Tiwana had stolen the ticket from Clarabel—the law would recognize Clarabel's true ownership.)

Legality, however, is not the same as morality. In this case, the tickets were earned only by students who raised funds for the charity. It wasn't like having a state lottery ticket bought by someone else and given to you as a gift. In this case, you personally had to take very specific actions—that is, raise money for charity—to be morally entitled to a ticket. A lawyer would say that since Clarabel had earned the ticket legitimately and given it freely to Tiwana, the ticket (and therefore the prize) belonged to Tiwana. A lawyer would not care about the moral offense to the other students who had worked to raise money for charity, who thereby earned entitlement to the prizes, and who therefore felt it was wrong for someone who had not raised one red cent to ride off on the grand prize. Mind you, a lesson about charitable deeds could also be learned by all those students (and their parents) who complained. What's the higher level of charity: helping those in need because you exercise compassion, or helping those in need because you want to win a prize? (Don't ask your lawyer that one either.)

Clarifying the moral point—that Tiwana had no moral entitlement to a ticket—allowed Ted to see clearly the course he wanted to take. He felt sure he could occupy the moral high ground and explain his reasoning to one and all, thereby defusing what had rapidly become a hot-button issue to a lot of people. He issued a statement that tickets were not meant to be transferable from one student to another—which included an apology for not making that clear at the outset. He explained his reasoning about rightful ownership of each ticket. He then announced that the ticket that won the bike (and therefore the bike itself) rightfully belonged to Clarabel. Clarabel would, of course, be free to keep the bike or make a gift of it to her friend. Note that no one could complain about Clarabel winning the bike: she had earned her ticket. On the other hand, no one could complain if she later gave the bike to Tiwana: it was hers to give. Whether Tiwana eventually got the bike isn't morally relevant: it's how she got it that counts.

In our society, many mistakenly think of the law as setting moral standards—anything legal, many suppose, is moral. Societies do reflect morals in their laws (for example, by criminalizing child abuse because it is harmful and therefore wrong), but societies don't necessarily become moral through legislation. Consider that genocide was legal in Nazi Germany, as were Stalin's purges (i.e., the murder of tens of millions of innocents) in the Soviet Union. Or think about abortion and capital punishment in America. Both are legal, but both have vociferous opponents declaring them immoral. Tobacco companies operate legally, but many view their entire business as immoral. On the flip side, many people privately believe in the morality of euthanasia, yet it remains illegal. If you learn nothing else from this point, it is that thoroughly thought-out ethical systems applied by reasonable people may still yield conflicts, because morality is not a subject like arithmetic: not all answers are objectively right or wrong.

The other key point highlighted by Ted's story is that good intentions are not enough to ensure that ethical standards are upheld. Remember the adage about the road to hell being paved with them (good intentions, not ethical standards!). Everyone involved with the fund-raising and the lottery meant well. But the conflict over the prizes was unforeseen, and its resolution required more work, and sleepless nights, than anyone would have guessed. However, once Ted's contemplation embraced the distinction between legal and moral entitlement, he was able to make his decision and regain his equilibrium as principal.

You are right to feel daunted by the task that lies in front of you. But the other lesson to learn from Ted is that you can work out an ethical solution to your problem. The Old Testament tells us that God saved Noah from the flood not because he was a paragon of virtue but because he was "blameless among the people of his time," or "righteous in his generation" (Genesis 6:9 and 7:1). Noah had his flaws, but he was apparently living in a thoroughly corrupt society. His efforts on behalf of the good life were enough to reserve his seat on the ark. You don't have to be perfect to be good.

JACKIE AND DAVID

Here's another case of someone trying to resolve a knotty moral quandary, to illustrate how you might manage competing personal interests.

Jackie and David moved to New York City from suburban California two years before their daughter was born. David had snagged a high-powered Wall Street job, and Jackie was more than happy to leave her corporate job for freelancing in order to experience life in the big city. And they loved every minute of their time in Manhattan—after they got over their sticker shock at the price of apartments—enjoying career successes, with growth that would have been impossible in their hometown, and the many cultural advantages of New York.

But now that they had a child, Tamara, and she was two years old, Jackie had started to worry about whether they should stay in the city. She was concerned about her daughter's safety and her education. As much as Jackie appreciated the city's opportunities, she wondered if the small California town where she had grown up and married would provide a more wholesome childhood for her daughter. She was familiar with the excellent public schools there and found the intense competition for slots at the top private schools in Manhattan—and the ferocity of competition within the schools—to be daunting and not entirely healthy. And the price! She feared she would never be able to let Tamara travel to school alone—or to the corner store, for that matter—in New York and treasured her memories of carefree afternoons outdoors with her childhood friends. For herself, Jackie didn't mind New Yorkers' relentless focus on achievement, but she was anxious about what impact the concentration on what you do and how much you make, as opposed to who you really are, would have on a growing girl.

On the other hand, as she knew from her own experience, the cultural and eventually the career opportunities available in Manhattan far outstripped those in suburbia. At private schools in New York, kindergartners learn chess and computing. Many of the graduating seniors are admitted to the Ivy League and are veterans of internships with leading research scientists. Jackie envisioned taking Tamara to a different art museum every Saturday, signing her up for voice lessons with a Broadway performer, giving her season tickets to the ballet for her birthday. Or in suburban California . . . running errands at the mall? Tap class at the rec center? Well, it had been good enough for her. But then, she'd had many restless weekends as a teenager.

Both Jackie's and David's parents and extended families lived on the West Coast. Visiting a couple of times a year was fine when they first moved, but now the aunts and uncles complained that Tamara was like a different human being each time they saw her, she was growing and changing so fast. Jackie and David left a large and established circle of friends behind in California, and many of those friends were now parents themselves. In New York they had many acquaintances they socialized with, but deep friendships are slow to develop. Jackie wondered if the advantages of New York could compensate for the distance from family and friends. One thing she knew for sure: if she was going to bother with the suburbs, she was going to do it near family and friends. No Westchester County for her.

For his part, David was content to stay in New York. But he too wanted to do the right thing for his daughter and told Jackie he'd move back to California if she was convinced it was crucial. They'd both pay a personal price, he knew, in their career advancement and their income, though he wouldn't be sorry to lower their mortgage payments and have a yard again.

The disparity in their levels of concern over the situation was causing tension in Jackie and David's marriage, even though they agreed that they wanted to do the right thing as parents. David wanted Jackie to get over her anxiety and relax and enjoy their family life—including their blossoming careers and bank account. Jackie wanted David to share her anxiety and work with her to find the solution, not just say he'd do what she decided.

Jackie had a long list of pluses and minuses about each situation in her head but couldn't seem to sort through them on her own. David had lost patience with the endless circle of doubts. Her best friend lived in California, so broaching the subject with her brought a sure response: come back home! That's when Jackie came to see me.

She faced a true dilemma: two choices, neither of which was entirely good. She felt strongly that she wanted to find and follow the moral path through this thicket—she wanted to be a moral person and a good mother. She knew intuitively that relationships were moral as well as emotional, and she wanted to do what was best for her child. She was committed to that, even if it meant choosing something less desirable to her personally. She was ready to put aside material things and career possibilities if necessary. She was, in short, trying to lead a good life. But she was running up against the inevitable question: what does that mean? What is a good life? What does it mean to be good? To do good? And good for whom?

Even if we could all agree on what goodness means (which, as we've seen, we can't), we would still have to ponder these questions carefully because the answers apply differently in different situations. A good banana, for example, is not the same as a good apple—the qualities making up *good* may change according to what you are considering. Similarly, it might mean one thing to be a good parent, another to be a good spouse, another to be a good employee. These interests often align, but when they don't mesh the conflict can be heartrending, as Jackie was finding out.

What finally propelled her to consult a third party was not her paralysis over making the New York versus California decision but her fear that the decision had potential to damage her marriage. What if the best thing for Tamara was to move back to California, but the best thing for the relationship between Jackie and David was to stay in New York? And could making a choice based on Tamara's needs at the expense of the marriage really be best for their daughter?

There was no doubt that Jackie had to make some decision, so the first thing I asked her to do was write down her list of pros and cons and categorize them. Rarely can you just tote up the columns and make your decision based on the results, but having something concrete to work with can help clarify your preferences and give you a basis for comparing them. For Jackie, it was a helpful first step in articulating what she wanted.

The remaining question was the big one: how do you weigh all these competing factors? The next phase was for Jackie to check her assumptions, so she would have a clear and true picture of her options. For example, Jackie said she was afraid of exposing her daughter to the violence of New York City. But the reality of violence in New York is at odds with the myth of the mean streets. The headlines are not the whole story, but even there she had overlooked the headlines about the plummeting crime rate while fixating on the headlines about the various senseless tragedies that make the news. Bad things happen everywhere, she had to realize, and relocating to the suburbs was by no means a guarantee of sheltering Tamara from harm. For that matter, Jackie should also weigh the vision of a sixteen-year-old Tamara by herself on the subway against the vision of a sixteen-year-old Tamara behind the wheel of a car, for the ultimate New York versus California tit-for-tat. Jackie and David would surely shield their child as much as possible no matter where they lived, giving her as good a chance as anybody of having a safe childhood.

With her array of options thus clarified, Jackie could see that there was clearly no one right answer, and even professional counseling wouldn't change that. The positive side of that was that neither situation was wholly bad. She'd have a mixed bag no matter what she did. She took comfort in the teaching of the Tao that you can't recognize the positive without some negative to contrast it with. The complications she faced threw into high relief the excellent aspects of each option available to her. Likewise, America's economic Golden Age, the tremendous prosperity of the Eisenhower and Kennedy years, was experienced under the shadow of the Cold War. We knew what we stood to lose, so we took pains to sustain it.

Jackie ultimately realized that with the loving care of her parents, Tamara would thrive on either coast, so the decision must be made on factors other than her needs alone. Jackie also knew that she would have to make her decision and then adjust to it, quieting her internal should-we-or-shouldn't-we debate in order to live fully in whatever situation she chose. With the relief of seeing that the decision was not as critical as she had imagined, she felt free to pick her advantages and disadvantages, given that no matter where she was she would be living in an imperfect world. There wasn't one absolutely right and one absolutely wrong choice. It was the fear that there was one but that she couldn't figure out what it was that had been gnawing at her. Finally, she realized that whatever she committed to would not, of course, be irrevocable. She had to give her choice a fair chance, but if she felt the flaws were finally too great to be outweighed by the benefits, she could change her mind.

Given that everything changes, the right answer now, with a toddler, might not be the right answer for the same family with a teenager, or with siblings, or with a new career for David. When it comes to the far future, philosophy isn't a tool for prognostication. It is of far more use in the present. And at present, Jackie understood that she was not in dire straits. She could be a good parent and give her child a proper upbringing on either coast. She understood that while there is more to making an ethical choice than flipping a coin, there is no guarantee that one option is absolutely better than the other. The right thing, then, is to get the best (and avoid the worst) from the choice you make.

Jackie, David, and Tamara stayed in New York. They liked it there, and the reasons they left California still applied. Jackie's anxiety about living in Manhattan with a child evaporated once she assured herself it was a moral path and that she wasn't somehow failing her child by satisfying herself and her husband. Her restored equanimity placed the marriage back on an even keel, so Jackie and David could focus on working together to provide a warm and solid home for Tamara.

MICHAEL

A colleague of mine, Keith Burkum, told me about one of his cases that highlights another aspect of ethical decision-making and moral living. Michael was the mayor of a small town (a part-time job). A religious organization planned to open a hospice in the town for people with AIDS but soon ran into vigorous local dissent. Though they had met all legal and zoning requirements, the hospice founders had not sought community input or public approval of their project. The protestors, who seemed to be operating solely on uneducated and irrational fears of HIV, demanded that Michael hold public meetings to air the issue, aimed at stopping the hospice from opening.

There were ethical and strategic problems on both sides of this conflict, but here we will focus on Michael's dilemma. He was caught between his responsibilities toward those who had elected him and his strong personal support of the hospice. He believed that the religious group was in fact working for the good of the community that wanted to block it—the same community that had elected him and now wanted him to put a stop to this new plan.

When Michael brought his story to a philosophical practitioner, they discussed Aristotle's ideas on the interplay of ethics and politics, which we tend to regard as entirely separate spheres, one private and the other public. For Aristotle, the good life is not just a matter of following a set of rules. Virtue ethics means developing character traits that will help you lead that kind of life. And it means considering not only what is best for you, but also what is best for the wider world you live in. He lists several virtues necessary for leading a good life, including courage, justice, temperance, and even a sense of humor.

Michael focused on justice. Aristotle wrote that for those in leadership positions, equity is a key component of justice, to be called on when a situation falls outside the scope of established rules and laws—as this situation did. Michael agreed with Aristotle's idea that politicians are responsible for the general ethical good of the community as well as the specific concerns of their constituency. So armed, Michael set out to find an equitable resolution to the conflict his town faced. He urged the hospice group, as members of the community themselves, to communicate more directly and openly about their program. He explained to the protestors that there were no health risks associated with the hospice and urged them to review the presented plans objectively, letting go of fears with no basis in reality in favor of something that could help so many people.

> "The equitable is both just and also better than the just in one sense. It is not better than the just in general, but better than the mistake due to the generality of the law. And this is the very nature of the equitable, a rectification of law where law falls short by reason of its universality."
> —ARISTOTLE

Michael felt this path followed the fine line he wanted to tread: a serious commitment to the wishes of the people he represented and the greater good of the community (where those two things didn't overlap) as well as to the dictates of his own conscience. Had this plan calmed the storm, he would have been satisfied that he had acted with equity. But he was mayor in the real world, and sometimes the real world doesn't behave just as philosophers think it should.

While the religious group reached out repeatedly to the community in general and the protestors in particular, their efforts were unrequited. Protest ultimately reached such a level that, fearing for his political future, Michael felt he had to allow the legal action demanded. The suit to shut down the hospice was ultimately defeated in court, so in the end the town got the hospice anyway.

Michael was optimistic that he could be reelected but was so discouraged about the role of virtue in a democracy that he was unsure whether he actually wanted the office for another term. Even though he was pleased with the final outcome, he felt he had compromised his principles in allowing the suit in the first place. On the other hand, even with hindsight he didn't see another, more satisfactory way through the battlefield.

Kudos to Michael for working so hard to act ethically in a world often hostile to those who do so. He can take teleological comfort in the fact that the outcome was the right thing, even if the process required to arrive at that place was not exactly the high road. Trying to clean up something dirty while staying clean yourself is never easy to do—and is often impossible, as Michael discovered. But where would we be if nobody ever tried?

THE MYTH OF GYGES RING

I'd like to leave you with one possible answer to a 2,500-year-old puzzle surrounding the difficulty of pronouncing something right or wrong. Plato retells the myth of the Gyges ring to lay the groundwork for his *Republic*. In a dialogue between Glaucon and Socrates, Glaucon relates the tale: A shepherd stumbles upon a magic ring known far and wide to cloak the owner in invisibility upon command. He loses no time in discovering all the things this power allows him to do—eavesdrop, steal, trespass—and in short order he has amassed wealth, seduced the queen, killed the king, and so become ruler of the land. He's undetectable whenever he wants to be, so he's gotten away with all this, and on top of that, he now enjoys the immunity of being king.

Could this be a case where crime pays? I don't think so, as moral misdeeds have a life of their own. If the Buddhists have it right (not that Buddhism was on Plato's agenda), the shepherd-king should take care not to sit with his back to the door. He may have gotten himself to the top of the mountain, but now there's nowhere to go but down.

But Plato's not done with this story yet. Glaucon asks Socrates how we can say that what the shepherd did is wrong. We'd all take the ring if given the chance, and if we had the ring, which of us wouldn't do the same things? Think of it: doing whatever you wanted with impunity. Socrates' (or Plato's) reply is *The Republic*, in which he describes a society so perfect that if a peddler came to town with a wheelbarrow full of Gyges rings, no one would pay a penny for them. If everyone had all they wanted, if everyone were satisfied and fulfilled, what good would a ring like that do anybody? Its only usefulness is in enabling you to get what you want but can't otherwise have.

Plato's optimistic position is that we should strive to bring about that better world rather than devoting our energy to devising ways to get away with things. Plato had a detailed political vision in mind for this utopia. But I think individual actions and personal responsibility are the first planks in the platform. You're laying the foundation right now by committing to work toward an ethical framework for your life.

"... of the unjust I say that the greater number, even though they escape in their youth, are found out at last and look foolish at the end of their course, and when they come to be old and miserable are flouted alike by stranger and citizen; they are beaten and then come those things unfit for ears polite. . . ."
—PLATO

Personally, I'd love to live in a world where people refrain from doing things because they are wrong, not just because they're afraid they'll get caught. I'm looking forward to the day when I see the cop on television look confused when a reporter asks him what went through his mind as he turned in the $35,000 rather than pocketing it. The money didn't belong to him. His job is law enforcement. What else would he do with the cash besides turn it in as evidence?

12

Finding Meaning
and Purpose

---

*"To declare that existence is absurd is to deny that it can ever be
given a meaning; to say that it is ambiguous is to assert that its
meaning is never fixed, that it must be constantly won."*
—SIMONE DE BEAUVOIR

*"Nothing contributes so much to tranquilize the mind as a steady
purpose—a point on which the soul may fix its intellectual eye."*
—MARY WOLLSTONECRAFT

A great philosophical plague of the twentieth century, sure to tail us
into the millennium, is widespread feelings of personal pointlessness.
So many people are without a firm sense of purpose or meaning in
their lives that the lack has come to seem normal. But few live happily
that way. We're generally not satisfied with the idea that our lives and
our world are completely accidental and without rhyme or reason. The
further we look in that direction without finding any other explanation, the harder it is to bear.

The existentialists are only partly to blame. They were so cool—
hanging out on the Left Bank, smoking cigarettes, thinking deep
thoughts, scribbling philosophy and poetry on napkins and table-
cloths. The existentialists truly excelled at making it look romantic to
kill off God and step into the abyss.

MAYBE IT'S JUST A PHASE

The exclusive focus on the dark side of the existentialists is our own fault for only scratching the surface of their work. At bottom, existentialism isn't just about angst and dread, or even ennui. When it is, it isn't good for you, because it robs you of much of the richness of life. That's why I regard existentialism as a phase—something you go through but get past. The most successful existentialists recovered a secular sense of meaning and duty from the ashes of a world formerly understood to be created and designed by a superior force. Existentialism asks, "Without God, without a grand design, what should we do?" Just following their path all the way to that end can restore your sense of purpose. As long as you're working with the assumption that there is a right thing to do, your purpose becomes to discover and do the right thing.

Existentialism also places a premium on authenticity, individual responsibility, and free will. So the good news is that you get to choose how to approach the void created by declaring God dead. A lot of people dip into existentialism, conclude that life is pointless, and wonder why, if that is so, they should bother with anything. Here's my favorite argument to stop that slide into existential depression: If life as we know it is indeed a fantastically unlikely accident, all the more reason to appreciate it. If we come from nothingness and will return to nothingness, I say let's spend the time we have celebrating the very somethingness of life. Our time here is precious—literally irreplaceable. So live authentically. The catch there is that you have to figure out what living authentically means to you, but one thing it surely implies is engagement with—not withdrawal from—life itself. Use your free will to choose renewed appreciation of every moment rather than despair.

These angles of existentialism are news to most people, however, and many are stuck with a vague notion of God being dead, hell being other people, the nausea of nothingness, and the absurdity of life. Not to worry. Your friendly neighborhood philosopher is here to help you peer through that darkness. Since I know you're going to ask: no, I do not have the final answer to the question "What is the meaning of life?" Even if I did, it might not be the same for you. Since this has been a classic line of inquiry for philosophers down through the ages, however, I do have some tools you can use to answer the question for yourself.

MEANING AND PURPOSE

The first key is to distinguish between meaning and purpose. Those words are generally used interchangeably, but I'd like to point out a distinction between them to help you reclaim them both in your own life. Purpose is an ultimate object or end to be attained. It is a goal. Meaning has to do with how you understand your life on an ongoing basis. Meaning is in the way things happen, not necessarily in the end result. Understanding depends on experience, and meaning—like experience—is very personal.

Say you're sitting in a restaurant, looking at a menu. What is the purpose of the menu? To help you choose something to eat. What is the meaning of the menu? To give you some information about your choices. If you're in a restaurant in France and you don't understand a word of French, then the menu will be meaningless to you—even though you know its purpose. So you can find purpose without meaning. On the other hand, if you go into a restaurant and understand the menu just fine but find the prices so outrageous that you can't or won't order any food, then the menu means something to you but serves no purpose for you. So you can find meaning without purpose. Then again, imagine someone who has never been in a restaurant before and moreover doesn't know how to read. The menu will have no meaning and serve no purpose for him. Finally, suppose the menu includes appetizing photos of the main dishes, and imagine someone who—instead of ordering the food—begins to eat the photos. That person would be confusing meaning with purpose.

The same thing applies when you take a trip in your car and consult your road map. The meaning of the map is its representation of the territory; the purpose of the map is to guide you to your destination. You don't normally suppose that by tracing your route on a map, you thereby complete your actual journey. Again, that would be confusing meaning with purpose. This is essentially the philosophical advice of Alfred Korzybski (and later Alan Watts), who cautioned that the menu is not the meal, the map is not the territory. Similarly, the meaning is not the purpose.

> *"A map is not the territory it represents, but, if correct, it has a similar structure to the territory, which accounts for its usefulness."*
> —ALFRED KORZYBSKI

So if you already have some purpose, then understanding the meanings of things can help you fulfill it. But if you have no purpose, or cannot discover one, then meanings are less useful to you. The most accurate map in the world is useless if you aren't going anywhere. Then again, you don't always want a map, or need to know where you're going. Shopping in a foreign city without a guidebook, or exploring a wilderness without a map, can be risky but also very rewarding. Your purpose can be simply to explore, and you can assign a meaning to everything you encounter on your journey. So we're back to individual philosophical dispositions: meaning and purpose depend a lot on you. Simple things can be very meaningful; inexplicable things can be highly purposive.

> *"I embrace the common, I explore and sit at the feet of the familiar, the low. Give me insight into today, and you may have the antique and future worlds. What would we really know the meaning of? The meal in the firkin; the milk in the pan; the ballad in the street; the news of the boat."*
> —RALPH WALDO EMERSON

We are much happier if we believe we have a purpose, independent of knowing what that purpose is or may be. But we are happier still if we know what that purpose is, for that helps us find meaning. Many meaningful things are not part of our purpose, though that makes them no less meaningful. We can also find meaning all around us without knowing what our purpose is (and so not knowing what fits into our purpose). You could also be sure about your prevailing purpose but fight meaninglessness every day. So purpose is no kind of guarantee for feeling your life is meaningful, in case you were thinking of simply signing on with someone offering you a ready-made purpose.

You might have one purpose that informs your entire life or, more likely, a shifting series of purposes at different times in your life. You might find parenthood your highest purpose for a long time. But by the time your children are grown, you may shift your emphasis toward your career, or personal development. Or, for another example, you might feel called to dentistry, but if you want to do more than endure retirement, you had better have another priority by the time you are not seeing patients every day. Your current purpose could also be to discover your next purpose, or your overall purpose, like the physiotherapist in Chapter 10 who left his job without knowing what he would do next to earn a living. Ecclesiastes is worth repeating here: "To every thing there is a season, and a time to every purpose under the heavens."

Purpose is more dogged than you might imagine, even if you don't recognize it. You should remember that if you ever consider adopting someone else's dreams for you as your own. It doesn't matter how long Aunt Millie has treasured her vision of you as a neurosurgeon, if playing the oboe is the only thing that moves you, go ahead and take the part-time job with the symphony and a day job to make the rent. Medical schools will always be there should you change your mind, and you will never rest easy if you ignore your own wishes. Your purpose won't be easily dissuaded.

I lead a monthly Philosopher's Forum at a local bookstore, and one of the regulars is a self-proclaimed nihilist—a believer in nothing, having no ideals, loyalties, or purpose. But there he is every month, always at the center of a knot of people trying to hash out some issue, clearly enjoying the reactions his extreme stance elicits. Could it be that his purpose is to tell everyone there is no purpose? Denying meaning is meaningful to him.

Without the flexibility to pursue different purposes during your life, you could end up like the "most popular" prom queen still living in past glory twenty years later without a new vision for herself. We are not made to do one thing. We are made to do one thing at a time. Don't cling to any one thing, stretching it beyond its time. If you have fulfilled a purpose, no one can take it away from you. But it won't last forever. Nothing does. You can relish it, and relive it, but you must be willing to relinquish it. You might not find another one if you can't let go of the purpose you have. It isn't easy to do, which is why you recognize the image of the prom queen who can't move on. It is a common condition. In "To An Athlete Dying Young," A. E. Housman writes "Smart lad, to slip betimes away / From fields where glory does not stay / . . . /Now you will not swell the rout / Of lads that wore their honors out. . . ." If you'd rather not die young just to avoid a major transition of purpose, your other option is to allow other purposes to emerge as you finish with old ones. It takes courage, but you have to go on.

Purpose isn't something you can get just because you are in the mood to have one. No one and no thing can give you purpose. You have to find it for yourself. True purpose may not be obvious and may take a long time to unfold, but that doesn't mean that it doesn't exist. As long as you find meaning along the way, that time is not wasted.

It's also easier to believe in an unknown purpose than to decipher an unknown meaning. Finding meaning can be an ongoing challenge. At the same time, you have to keep things in perspective. While you're stuck in a monster traffic jam, your frustration and the escaping minutes can obliterate meaning and purpose alike. Rather than developing road rage, you'd do better to contemplate the inevitable passage of time, and your best use of it.

"Hold fast the time! Guard it, watch over it, every hour, every minute! Unregarded it slips away. . . . Hold every moment sacred. Give each clarity and meaning, each the weight of thine awareness, each its true and due fulfillment."
—THOMAS MANN

Instead of honking your horn, or cursing the imbecile who just cut you off, or getting into a gunfight with him, take a deep breath. Instead of giving the finger to someone who honks at you, consider how fortunate you are to be breathing. You can't dissolve the traffic jam, but you can dissolve your stress at being caught in one.

In the poem "If," Kipling's suggestion for successful living is to "fill the unforgiving minute with sixty-seconds' worth of distance run." He was driving at finding meaning in the small moments of ordinary life, rather than just wasting them, and the emergence of purpose out of the accumulation of those moments. That road is paved with meaning, and it leads to purpose.

> *"If you can fill the unforgiving minute*
> *With sixty-seconds' worth of distance run—*
> *Yours is the Earth and everything that's in it,*
> *And—which is more—you'll be a Man my son!"*
> —RUDYARD KIPLING

Equivalently, you'll be a Woman, my daughter (only that doesn't scan or rhyme).

It can be very difficult to understand our lives at times. We'd like to find a pattern—something more than an accrual of habits, or a drive to pass on our genes. We want that pattern to be pushing us toward better things. Such an optimistic outlook is the gentlest salve when something pains us. Bad things may happen, but at least experiencing them can help make us better people. Philosophers from Heraclitus to Lao Tzu agree that change is the only constant in life, and we all have our ups and downs. We'd like to believe that what we go through (particularly the bad stuff) is meant to teach us a lesson and to allow us to be more than we otherwise might have been. I can't tell you whether anyone or anything has given us these experiences with any intention at all about what if anything we learn from them. But as individuals with freedom of will, we can certainly choose to use anything that comes our way as fodder for our personal evolution.

HINDSIGHT IS 20/20

Many of my clients find it helpful to talk and reflect about meaning and purpose. Those who feel lost often realize that they actually have a line on one, which anchors them as they discover the other. Realize that you don't have to be able to put your finger on both at every minute to have a fulfilling life. Current uncertainty about your purpose is not the same as having a purposeless life. Furthermore, it is a mistake to assume that your experience is without meaning just because you don't immediately see its meaningfulness. I'll demonstrate it to you.

Consider a man who, looking back on his life after sixty-six years, thought he was an abject failure: Winston Churchill. He'd been a soldier, journalist, parliamentarian, and author, publishing his first book before he was twenty-five. He had served as First Lord of the Admiralty. He was elected to Parliament while still a young man and subsequently held a wide variety of powerful government positions. But he was convinced that his real task in life was to be prime minister and, having failed to attain that office thus far, he felt worthless despite his many accomplishments. His time came, of course, and history holds a view of the man completely contrary to his own opinion of himself at sixty-six.

Hindsight being 20/20, we can see that all his earlier experiences were necessary preparations for his government's highest office, and that he wouldn't have been the great world leader he was if he had moved to 10 Downing Street too soon. Fatalists would say that history had reserved his ultimate purpose—to stand up to Hitler and win the Battle of Britain—although he couldn't have known that while preparing to fulfill it.

You don't jump from nowhere to somewhere. You are always somewhere. Even if you don't want to be where you are right now, or don't know where you are, it is still a point on your path. Churchill thought he had missed his calling, without realizing that he was still on the right path only not yet at his destination. If you, like Churchill (enjoy this now—it isn't every day you get to say, "Oh, yes, I'm just like

Winston Churchill"), think you are lost, it may be that you just don't see the pattern yet. You may be headed toward something—or already engaged in something—important for you, without knowing it.

A meaningful life always has an intricate pattern, and if it is going to be intricate, it will have elements we don't understand as they are happening. Only later will we see where they fit in the pattern. Eudora Welty recommends "an abiding respect for the unknown in a human lifetime and a sense of where to look for the threads, how to follow, how to connect, find in the thick of the tangle what clear line persists. The strands are all there: to the memory nothing is ever really lost."

GOD NEEDS YOU

The early Greeks saw the world as an orderly place, with everything unfolding for a particular purpose, or "final end," called *telos*. Their philosophy of purpose is called "teleology." They assumed that people, too, had a purpose, just as all things in nature did. That way of thinking was reflected by Jewish scholars, and incorporated by Christian thinkers. Theologically, the purpose of earthly life is to prepare for heaven, or get ready for the Messiah or for Judgment Day, or to redeem your soul, or something similar (depending on your particular theology). Religion has been so successful through the ages in part because it provides meaning and purpose to individuals. Those things get harder to come by in a consumer-driven society—you can't find meaning or purpose in a Sears catalog or order them from the Home Shopping Channel—so the clear path of an organized religion can be greatly appealing.

"How would man exist if God did not need him, and how would you exist? You need God in order to be, and God needs you—for that is the meaning of your life."
—MARTIN BUBER

Judeo-Christian traditions assume that there is a higher meaning to life. But as our society has moved further and further from those roots, just what that meaning is has been obscured. If you're still drawing nourishment from those roots, you're ahead of the game when it comes to knowing your purpose. All highly structured institutions—like religions, or the military, or large corporate employers—provide you with meaning and purpose in return for all they demand from you and the order they impose on your life. But no one, including a God, a general, or a CEO, gives you a complete, ready-made meaning or purpose, signed, sealed, and delivered. It is more like providing you with the clay that you must sculpt. But at least you're not trying to figure out how to make the clay. For those who are, this chapter is as close to a recipe as you're likely to find.

THE MONK

Just in case you're wishing you had a religion to do part of this work for you, here's a case to assure you that faith offers no lasting guarantees. A philosophical counseling colleague, Ben Mijuskovic, had a client, Fred, who had been a monk for ten years. Fred had been struggling with the symptoms of depression—fatigue, insomnia, hopelessness, helplessness, and even thoughts of suicide—for some time. When pastoral counseling didn't relieve his suffering, he visited a psychologist. He'd had a happy and secure childhood and had been happily devoted to his order for most of his adulthood, so exploration of his past brought no relief either. He ultimately tried medication, with no success. After talking with a philosophical counselor, he came to see why.

Fred explained that the most painful part of his depression was that his faith was no longer meaningful to him. As he continued, he also bemoaned what his vows had cost him: he had no biological family, no sexual intimacy, no regular participation in the everyday world. For nearly all of his years in the monastery, he'd gratefully accepted the benefits those sacrifices brought: a deep spirituality, a personal relationship with God, and the ability to share the abiding peace and joy those brought him with others. He hated his depression for robbing him of the satisfaction his monastic life had once given him.

Ben asked him to consider whether it was possible that the loss of meaning could be causing his feelings of depression, rather than the other way around. Americans have been bombarded with so much propaganda about chemical imbalances causing depression that we've lost sight of the fact that our states of mind may have a crucial say in our brain chemistry too. Not all depressions have strictly physical origins. Ben wondered if Fred's might not be philosophical. After all, he'd already tried a medical approach without success, suggesting that something more than an imbalance of chemicals was at work.

Fred barely paused before answering. As soon as he heard the question, he realized that he'd simply been looking through the wrong end of the telescope. He'd been so thoroughly (and, until recently, happily) entrenched in being a monk that he didn't recognize the shift in his beliefs when it started. As the sessions continued and Fred had time to digest this new knowledge of himself, he admitted that he was no longer satisfied with a cloistered life. He'd grown, matured—changed. The path he was on was no longer the way to his true self, he feared. He had lost his sense of purpose, and along with it, his sense of meaning in everyday life. Fred had never experienced adult life without a definitive purpose. No wonder he felt depressed!

Eventually Fred decided to leave the monastery. Renouncing religious vows is not to be undertaken lightly, of course, and it was a painful decision. But even as he struggled to build a new life in the wider community, Fred's depression evaporated. He described a powerful sense of liberation and renewal. He remained a deeply religious person, and even brought some of the rituals of the monastery with him into his new life. He was unsure about what new purpose—or new phase of the same general purpose—would illuminate his life, but was willing to wait for that to develop.

Finding purpose and meaning can entail a lot of work, even given religious faith. Fred's story also demonstrates that anyone can come to a philosophical crisis—and work through it successfully, no matter how profound or complex. As with Fred, the philosophical insight that turns you around may be small. The power comes from your taking the time to absorb its full impact.

GOD MUGGED ME

Another of my colleagues, Peter Raabe, had a client with a very different spiritual conflict, who also resolved the issue philosophically. Sherman was a recovering alcoholic and drug addict who had spent his teenage years robbing and stealing to pay for his habits. Several years before, a sudden awakening to the complete waste of such a life motivated him to get straight, enroll in college, and begin working part-time.

Sherman was Native American but had been adopted as a baby by a white Christian couple. An integral part of turning his life around had been seeking out a spirituality informed primarily by his biological ancestors. He learned about a transcendent Spirit honored by his tribe and combined that belief with complementary New Age ideas about God. His personal faith in a loving and benevolent God in charge of everything allowed Sherman to forgive himself for his past lifestyle— and to feel forgiven by God's grace. He believed that everything was a part of God's plan and understood everything that happened to him to be a result of God's love.

But Sherman was thrown into a crisis of faith when he was mugged by knife-wielding assailants one night on his way home from the movies with a friend. How could God allow that to happen? Especially now, after his hard-won success in cleaning up his act? Was this some kind of divine retribution for his earlier sins? Or punishment for more recent sins—even though nothing he'd done of late could compare with his past history? He got angry at God for betraying his trust and started to wonder if he'd been wrong about God all along. Then he felt guilty about being angry at God and about questioning his beliefs. Then he got even angrier over being made to feel guilty. For Sherman, the mugging itself was not as bad as the way it shook the foundations of his faith. For if those foundations fell, he'd lose his identity as a loved child of a beneficent God. What would his purpose be if not to serve God's purpose? There'd be nothing left between him and a life of crime.

Sherman was experiencing the conflict between his assumptions (a benevolent God controls everything that happens to us) and his experience (I was mugged). Since the experience itself was undeniable, but he couldn't just drop his beliefs, he sought philosophical counseling.

Sherman was trapped by his own logic. As difficult as it was for someone who considered himself full of faith, Sherman began to check his assumptions. With his philosophical practitioner, he listed some alternative explanations: God does not plan everything. I'm being tested as God tested Job. God is sometimes wrathful. God doesn't exist. I didn't do anything wrong. I wasn't sufficiently vigilant about personal safety. God does not control everything. This is designed to increase my empathy toward the many kinds of suffering in the world. I have left too much in God's hands and not taken enough personal responsibility. The mugging was just an accident, nothing personal. The only ones responsible for this mugging are the muggers themselves.

It was this last idea that released Sherman. He knew from his own days on the handle end of the knife that God didn't figure in the muggers' plans. They were surely concerned with finding likely looking victims (smaller stature, physically isolated, affluent appearance), not with whether this was what God wanted them to do. Sherman knew that a benevolent father-figure, God, wouldn't be keeping a running tally of past sins and planning retaliation, especially when the person in question had made such strides. What Sherman had been unable to see while fixed on the idea that the mugging was part of God's plan for him was that perhaps neither he nor God were to blame. If anyone should be called to account, it is only the muggers themselves.

Sherman's distress turned out to be useful in that it spurred him to closer self-examination and to seek some resolution of his problem. He realized that it is pointless to be angry at God. And though it would be natural to be angry at the muggers, that wouldn't get him anywhere either. He had chosen to direct his anger at himself for having assumed that everything would always be good and for having shirked some personal responsibility on the pretext of leaving everything in God's hands. But he refused to wallow in that anger and sought to use it as an entree to constructive contemplation and change.

Having the courage to consider that some beliefs guiding his life might be false gave Sherman the confidence to reclaim them, albeit in an expanded form. Philosophical discussion helped him reach a place where his beliefs and experiences were consistent with each other. He modified his worldview to include the fact that sometimes bad things happen to you, and they happen not necessarily because you are bad. Sherman relieved God of micromanagerial duties without giving up his faith in a guiding force of essential goodness. He used his crisis to examine and deepen his spiritual commitment.

WITHOUT GOD, IS THERE PURPOSE?

It is not necessarily the case that if there is no God, there is no purpose. You don't need to despair just because you aren't sure about the existence of God. If Genesis doesn't satisfy you as the explanation of life as we know it, that doesn't mean there isn't an explanation. Many plausible theories exist, as you can see by the length of the list Sherman constructed just to address his specific situation. Even if it were all an accident, there would be no reason to believe that your life is without purpose. If you won the lottery accidentally, you would see all kinds of purposes for that windfall. Accident can bequeath important things, the great gift of life as we know it being, perhaps, one of them. The only way to do it justice is to live as fully as possible. Too often we don't.

Trials and tragedies often set us on a course for discovering—or rediscovering—our purpose. That is one way to find meaning in the most difficult times. We need a belief in the order of things to make sense of our world, to reach the kind of understanding that is required for meaning. Every culture has connected the dots of stars in the sky into different constellations, projecting order onto randomness as a way of comprehending it. When we can perceive a pattern, we have meaning. When we have meaning, we can find purpose.

We tend to reject unpleasant things as having no place within the pattern, but some philosophies, like the Tao, always account for the intertwining of opposites. If you seek good, you will also encounter evil. If you seek meaning, some inexplicable things will happen to you. It's likely that if you don't understand an event as part of the pattern, it's because you haven't seen the whole design yet.

MARTINE

Martine also struggled to coordinate her purpose with her experience. She worked at any job she could get that was related to film production, but what kept her going through the gofering and location scouting was her vision for her own film. It would be ecological. Urgent. About saving the planet. It would portray the future we're headed toward so starkly that everyone would have to take note and reexamine their own impact on the earth. In her favorite fantasy, audiences all over the world would unite to seek solutions. She laid her plans and meanwhile slogged through the thankless on-set dirty work. (You've been reading carefully, so you can see that Martine had a clear vision of her purpose but found most of her day-to-day experiences meaningless.) In her free time she volunteered with a grass-roots environmental group, and while that work was very important to her, the more she learned about overpopulation, global warming, and pervasive pollution, the more despairing she became.

Years went by, and though Martine had by this time built an extensive knowledge base in both movie-making and environmentalism, her own movie somehow never got off the drawing board. It was becoming clear to her that humans would most likely be extinct in another thirty years, so what would be the point of the gargantuan effort required to make a movie? This deflation of her dream left her feeling as if her life had no purpose. She became despondent and dysfunctional; her most important work, by her own admission, was being left undone. Her paralysis worried her, and so she found her way to my office.

The Taoist sage Chuang Tzu wrote that the wise person avoids disaster by regarding what appears inevitable as preventable. His companion insight is that the unwise person courts disaster by regarding what is preventable as inevitable. I discussed this with Martine and challenged her to consider how she could possibly know whether the future she saw was truly inevitable.

> *"The sage looks at the inevitable and decides that it is not inevitable.*
> *. . . The common man looks at what is not inevitable and decides*
> *that it is inevitable. . . ."*
> —CHUANG TZU

I also asked her to put her situation in a broader perspective. To that end, I pointed out that in a few billion years, our sun will become a supernova and its expanding fiery envelope will incinerate the earth. Should this impending catastrophe send us running for sackcloth and ashes, lamenting the late great planet Earth? No, she said, that was too far off to worry about now. I suggested that if she wasn't demoralized by the certain destruction of the planet in the far future, she shouldn't be depressed by the possible extinction of our species in the near future. Who knew what would happen in the coming decades? She countered that the trends that alarmed her were already far out of control; she felt it was too late to make any difference.

In that case, I said, she could take a tip from Chuang Tzu's logic. If irresistible forces have already fated the human race to become extinct, then there is nothing you or anyone can do about it. So you should go ahead and make your movie, I told her. It will give you a sense of purpose and certainly can't do any harm. (She didn't seem to have any other big plans for the next thirty years.) But consider at least the possibility that our fate is not already sealed and that we have the means to prevent our own extinction. If that is so, you should make your movie, I told her, for it may help. Either way: make the movie.

I also advised Martine on a purely practical level, suggesting that she break the making of the movie down into a series of manageable goals. Anyone might be daunted if confronted all at once with the necessity of raising all the necessary cash, rounding up all the necessary people, and filling all that screen time. Taking it one step at a time, on a purely practical basis, would help get her going and keep her going. It was clear that creating that movie was central to maintaining a sense of meaning and purpose in her life, so it made sense to arrange the logistics to facilitate the process as much as possible.

A big thing can be simply the sum of small things. A life of purpose is built step by step. A meaningless life does not usually fill with meaning in a blinding flash of light—few experience that kind of epiphany. You can manufacture it for yourself, but in small pieces that you must assemble gradually. Don't wait for outside forces to do it for you, because that day may never come.

Fulfillment of your larger purpose may lie in the future, but the practices that get you there happen today. Squirrels spend autumn storing nuts for winter. Even though they won't be needed for months, each nut counts. Try this experiment: set a modest goal, and attain it. See if a sense of accomplishment doesn't make you feel better already. Clean the house—or just one closet. Audit a course. Learn self-defense. Take up bridge. Anything within your means will do. It may not always be so literal, but a succession of short-term purposes can add up to a long-term purpose. Purpose, like meaning, often appears retroactively.

Martine's view of the future was debilitating her in the present. She was creating a self-fulfilling prophecy by allowing her fears to paralyze her. She made a logical error in assuming that because there was no future, there was no present. On the evidence that we were sitting there engaged in dialogue, there was in fact a present, and since it was undoubtedly there, it had to be reckoned with. It is important not to dwell too much in the past or the future, because they only obscure the present, which you must contend with in any case. For Martine, accepting her present was enough to restore her sense of daily meaning and renew her purpose to make her movie a reality.

MARTHA

The cases I've presented so far had to do with crises of purpose. But a crisis of meaning can also be addressed by philosophical counseling. My British colleague Simon du Plock counseled a young woman who was studying abroad for a year. After a stellar academic performance early in the first semester, Martha began to miss classes and to remain silent and withdrawn when she did attend class. She confided that she felt she could no longer handle the work and was terrified of failing her final exams.

The issue at the crux of Martha's rapid downward slide wasn't immediately obvious. But in counseling, she revealed her core problem: she had elected to share an off-campus house rather than live in a dorm and felt that her housemates were taking advantage of her financially. She felt they treated her as a privileged foreigner and refused to relate to her personally. She had dreamed of experiencing the "real" London by living with Londoners and felt sorely deprived by failing to establish friendships with them because of the wall they erected between themselves and her.

Let's continue looking at this case in the context of the PEACE process. Martha's emotions included anger, fear, disappointment, and rejection. She also sensed injustice and felt a lack of self-confidence. Like most clients, she was well aware of her emotions before beginning philosophical counseling. But she had done less analysis than many people do on their own. She saw herself as a victim, the unwitting dupe of ill-intentioned strangers, alone in a world she couldn't make sense of, which didn't respond to her actions in the ways she had anticipated. In talking with her counselor, she connected her surface issues with her experience of being far from home, living independently of her parents, and traveling abroad for the first time. Setting the course of her life fully on her own for the first time had turned out to be more complicated, more difficult, and less successful than she had envisioned.

A philosophical insight in the contemplation stage calmed the seas Martha was sailing on, and it boiled down to the meaning Martha took from her experiences—in du Plock's words, "the clarification of the meanings embedded in the client's language." When asked why she came to London, Martha said she had wanted to improve her education. Asked to define *education,* she talked about grades and degrees and doorways to well-paid professions—an outcome-oriented view of education as a means to an end. Her philosophical practitioner suggested a broader way of looking at education as experience—and what she learned from experience. Looking at it that way brightened Martha's outlook immediately. From that perspective, she took pride in her adventurousness, even if her adventures hadn't turn out as she'd planned—or as she would have liked. She began to see her efforts to establish an independent life as an important rite of passage from child to adult. She also took responsibility for making bad decisions. That was less comfortable in the immediate moment than playing the helpless victim but in the long run restored her sense of her own power. If she could bring about a negative situation with injudicious actions, she could also bring about a positive situation with more cautiously considered plans next time around.

Implementing what she had learned restored Martha to a state of equilibrium. She chalked up her unpleasant housing situation to a cost of her education in life—what she was here for, after all—and sought a change of venue for her second semester. With her mind cleared, she refocused her energies on her academic studies in time to buckle down for her finals. Clarifying the meaning of her life experiences, and connecting them to a purpose that resonated with her, afforded Martha the peace of mind she needed to succeed.

GRADUATION

As long as we are in school, we are used to a series of progressive units leading us toward greater goals. Each year you enter a new level, climbing another rung toward graduation. The purpose of third grade is to get you to fourth grade, but the meaning of third grade comes from what you're learning and doing each day—usually with no thought of fourth grade. So you learn how to read chapter books and do long division, how to lay out the rules for a game of kickball at recess, what happens when you bring home a mediocre report card, and how to make and keep friends. But even that third grade progress toward fourth grade is serving a still larger goal: preparing for high school and perhaps higher education.

In life, as in school, when you learn what you need to know at one level, you graduate to the next. But life isn't as neatly sliced as school, of course. No one is giving you summer vacations or handing you caps and gowns to mark graduations. You have to recognize the transitions yourself, and doing so is the key to lasting fulfillment. Otherwise you'll still be in third grade while everyone else is getting driver's licenses and moving into their own apartments.

If you believe in reincarnation, life is just like school. Each lifetime is lived in preparation for the next, until eventually you graduate from the School of Life and Death. But even if transmigration of souls isn't your cup of tea, that sense of evolution to progressively advanced levels should stand you in good stead.

Second Mother

One of my college mentors was a scientist who had risen to a senior professorship at a time when very few women went further than high school biology. In a competitive and then largely male world, Irene devoted all her energy to her career and was rewarded with increasing recognition of her excellent scholarship and brilliant research. This came at a personal cost: though she enjoyed a strong marriage, Irene had never felt she could step back from science long enough to bear and raise children. As she grew older, and her standing in the academic world solidified, she realized that she was sorry to have missed the experience of motherhood.

And so, after long (and successful) devotion to one purpose, a second, surprising one emerged. Starting a biological family was no longer an option for Irene, but neither was floundering in regret. Instead she created a program to introduce promising students—young women in particular—to science careers beginning in their freshman year. It wasn't the kind of thing senior faculty usually worked on, but it was soon a resounding success, and a number of today's scientists look back and credit her with igniting and encouraging their interest.

The success of any endeavor she undertook would surprise no one, but the mother hen role she adopted with the students no one would have foreseen—including herself. Irene found a new purpose in being a surrogate mother to these groups of young men and women, as well as in being an intellectual mentor. By fearlessly recognizing shifts in her sense of purpose, Irene combined and fulfilled two important aspects of her personality, never losing her sense of purpose while expanding her sources of meaning.

RED LIGHT, GREEN LIGHT

After all this, you may be ready to give up working out your purpose in favor of Destiny. Fatalism can be destructive, as we saw in Martine's and Sherman's cases, but it can be appealing in that it provides closure of a sort. You can wash your hands of the whole matter, humming "Que Sera, Sera (Whatever Will Be, Will Be)" and giving your very best Doris Day impersonation. But in the end, fatalism robs you of purpose. You're just acting out someone else's script. You become a passive agent, not responsible for anything.

Tolstoy was a big-time fatalist. According to Tolstoy, Napoleon was not at all responsible for the carnage of the Napoleonic Wars—he was just a pawn in a cosmic game; all his moves were ordained at the inception of the universe. Personally, I don't think that's healthy. We don't know anything about fate. We don't know whether human history is an unfolding of destiny, a question of chance, or a matter of will. We've got little certitude one way or the other, and without it I subscribe to what's best for our moral characters. It may require more work on our part, but I believe in free will and in being held accountable for our actions.

"Kings are history's slaves."
—LEO TOLSTOY

Isaac Bashevis Singer said, "When I'm crossing the street with a fatalist, I notice something strange: The fatalist usually waits for the light to turn green." In case you miss the point of Singer's philosophical joke, if you believe that everything is predetermined, then you also believe that the exact time and cause of your death are predetermined. Since there's nothing you can do to change that, you can't be killed before your time—even if you jaywalk in Manhattan. Unless, of course, you are fated to die by jaywalking in Manhattan. (The quick-thinking fatalist's reply to Singer would be "I was fated to stop at the red light.")

DEPRESSION

There is a place for philosophical considerations even in cases where depression is biological in origin, not just existential. Existential malaise, or death of spirit, may be compounding your situation, in which case philosophical work will help at least in part. The lack of purpose, animation, and excitement about life that characterizes depression can be addressed by philosophy in addition to psychology and medicine.

If you are depressed, whether or not there is a biological component, you most likely feel robbed of meaning or purpose. Consider this: it may be that your purpose is to get beyond your depression. The same holds for anxiety, grief, or general unhappiness. Your affliction doesn't just detract from your quality of life; it also gives you a challenge to overcome. Even with conditions considered purely biological, like cancer, successful treatment has been shown to rely not just on medicine but also on attitude. Your outlook and your disposition—your philosophy—can influence the outcome of the battle. Those whose purpose lies in winning the battle have a better chance of doing so.

BOREDOM

The most common cause of loss of purpose is simply boredom. Boredom appears to be almost unique to humans—no other animals, in their natural habitats, exhibit boredom. They are basically too busy for it. They are looking for food, trying not to be food, defending their territories, finding mates, rearing their young, preparing for the next season. This holds true for wild animals and even domesticated animals that depend on humans for survival. Captive animals (i.e., wild animals caged in zoos) do exhibit boredom, along with abnormal or demented behavior, which is worth noting since it is clearly a condition of the captivity, not the animal.

The human animal falls a bit into each of these categories. Sometimes we are wild. More often we are domesticated (or if you prefer, civilized). You might hear someone say, "Marriage has domesticated him," though that's often a complaint (made by, say, former fraternity brothers against a "lost" colleague). We may resist it, but we need some degree of domestication to get by, both individually and socially. We are also captive creatures, though not usually literally. More figuratively, we are captives within the limits of our language, culture, and experience.

No one—human or other animal—is bored in a crisis. In a catastrophe, everyone has a purpose: preserving themselves. Nonhuman animals exert less control over their environments than humans do, so they face more daily threats. We've thought and fought ourselves onto a relatively easy street, so for most people in developed nations, life is not a long series of struggles for enough to eat. But there is a danger in having everything you need—and even more danger in having everything you want. If material things are your goal, and you've achieved that particular end, the feeling of having no more mountains left to climb is dispiriting. Alexander the Great supposedly wept because he had no more worlds to conquer.

I'm not recommending creating catastrophes just to keep yourself interested. Mild thrill-seeking (like roller coasters) or even "extreme" sports are enjoyable because they invoke the feeling of encountering danger in a nonlife-threatening setting. More risky adventure fights boredom too but takes you far from the "middle way" that Buddha and Aristotle recommend. Those who climb Mount Everest may not be bored as they focus on their singular purpose—standing twenty-eight thousand feet above sea level—and fight just to keep breathing. But as the tragedy chronicled by *Into Thin Air* demonstrates, when the worst does happen the purpose that had seemed so sweet at the outset sours quickly. Even for those who come back down, what is next? For too many it seems to be a long series of climbing peaks despite increasingly excruciating risks. Does it really require regular brushes with death just to feel alive? Not for those willing to climb internal mountains through philosophical work.

There is one very simple strategy we can develop from understanding the impact of boredom on our feelings of purposelessness. Some of us need a trip back to the wild to refresh our sensibilities. When domestication—and captivity—have done their worst, a breath of fresh air can clear your head. "Back to nature" will mean different things to different people. Don't stress yourself out by camping where there's no running water if you're the type of person who is miserable without a daily shower. Find what is right for you: a walk around the block on a sunny day, puttering in your garden, or a month-long solo sojourn in the wildest part of a national park. Whatever reconnects you to the natural world is the surest way to regain perspective on your life as just a part of a greater whole: neither an isolated, incidental part of a massive hunk of chaos nor a pathetic cog in some relentless machine, but an integral part of a complex and vibrant system. This is what Thoreau discovered at Walden Pond.

> *"Undoubtedly the very tedium and ennui which presume to have exhausted the variety and joys of life are as old as Adam. But man's capacities have never been measured; nor are we to judge what he can do by any precedents, so little has been tried."*
> —HENRY DAVID THOREAU

Experience with nature helps rekindle an appreciation of life itself, one of the best ways of finding meaning and purpose. Understanding that life is a great gift, and enjoying all the things you do as part of your daily life, are also great antidotes to purposelessness.

HELP YOURSELF BY HELPING SOMEONE ELSE

The surest way to counteract feelings of emptiness in your life is to help someone else. It gives you a meaning and purpose you can't deny. It may help you see opportunities in your life to which you were formerly blind, and seeing into another's world helps you escape the captivity of your own. Feeling that life is meaningless is in a sense a luxury. If you were scrambling to hold body and soul together, you'd never pause to question the meaningfulness of your actions. So if you've read this far, I'm betting you're one of the affluent. Connecting with someone less fortunate than you are should be a powerful lesson to you. At the very least, you'll feel grateful for what you have.

> *"If you contribute to other people's happiness, you will find the true good, the true meaning of life."*
> —DALAI LAMA

GRIN AND BEAR IT

Your final simple option when struggling with meaninglessness or purposelessness is to bite the bullet: grit your teeth and hang on. Things change. Once you're at rock bottom, things can only get better. (If they get worse, you obviously weren't at rock bottom yet!) Herman Hesse won the Nobel Prize in Literature, but earlier in his life he seriously contemplated suicide, so deep was his conviction that there was no point to living. His talent as a writer blossomed later in life, but in the meantime he couldn't see his purpose and, without one, the meaning of his everyday life. His books later explored problems of personal identity, inner meanings and hidden purposes of life, and patterns on the path to enlightenment. The difficulty of his own road illuminated his writing, which in turn inspired a generation, although that was a purpose surely lost on him as he struggled in his early years. Ultimately he toughed it out until things got better. It is up to you how much consolation you find in the notion that if you can exercise patience and courage—two cardinal virtues—change will come. We are almost always able to extract meaning and purpose from events, even horrific events, but sometimes that takes time.

13

Gaining from Loss

"If a man in the morning hears the right way,
he may die in the evening without regret."
—CONFUCIUS

"Every old man knows that he will die soon. But what does
knowing mean in his case? . . . The truth of the matter is that the
idea of death's coming closer is mistaken. Death is neither near
nor far. . . . It is not correct to speak of a relationship with death:
the fact is that the old man, like all other men, has a relationship
with life and with nothing else."
—SIMONE DE BEAUVOIR

Consider the Buddhist parable of the mustard seed. A distraught
young mother mourning the death of her infant seeks Buddha's counsel. She says she is terribly sad and unable to get over this devastating
loss. Buddha tells her to go to every house in her village, collect a mustard seed from each house that has not known death, and bring all the
seeds back to him. She diligently goes door-to-door, and as she leaves
each one empty-handed, she realizes that there is no home untouched
by death. She returns to Buddha with no mustard seeds, and he tells
her what she has already seen: she is not alone. Death is something that
happens to all of us, to every family. It is only a matter of time. What is
inevitable, he tells her, should not be lamented to excess.

This insight doesn't bring the mourned person back, but it does raise
one's awareness of death as a necessary part of life. When death does
come, it needn't be a catastrophe that overtakes the living. That won't

erase your grief, or your need for mourning, but it should help you face it more stoically—philosophically—or at least without surprise or shock.

> *"For certain is death for the born*
> *And certain is birth for the dead;*
> *Therefore over the inevitable*
> *Thou shouldst not grieve."*
> —BHAGAVAD GITA

A BIG SURPRISE

In the West, we are always shocked by death. We don't deal well with it. In fact, we hardly deal with it at all. We like it well enough as entertainment, gorging on violent television, movie, and video-game deaths. Take it off the screen and into real life, however, and we can't bear to watch. So we cruise along on the good ship *Denial,* thinking life goes on forever, thinking death won't happen or won't happen to us. We conceive of death as the worst thing and therefore don't want anything to do with it. We've designated hospitals and mortuaries to do our dirty work, so we don't have to deal with it—until we end up on the doorstep of one of these dreaded institutions.

The ability to turn away from the reality of death is a modern luxury. It wasn't that long ago that death had a place in ordinary life. Generations of a family lived together, and people were born and died at home. Conditions that are now rarely fatal routinely killed people in the days before antibiotics and other medical advances. Parents did not expect to see all their offspring live to adulthood; as many as half their children died in childhood. Life spans were significantly shorter. When you died, you might have been laid out in your own living room. Death was commonplace, expected, tangible.

Now the death of a loved one, or the prospect of our own death, becomes an unbearable burden because we are completely unprepared.

If nothing else, death is a natural part of the life cycle, but any energy we have left over from denying death we put into staving it off. There's very little left for accepting death as inevitable. Maybe it's because we are biological organisms that will do anything to stay alive. When its leg is caught in a trap, wolves and other animals have been known to gnaw off the leg in order to remain alive, if three-legged. We have an instinct for self-preservation. Freud called it Eros, the instinct (or natural drive) for life itself.

But thinking death won't happen, and then being devastated when it does, isn't productive. The first big hurdle you face in handling the loss of a loved one, or confronting your own mortality, is to acknowledge death as a part of life. Being prepared in this way does not mean the loss won't cause you pain. But from the acknowledgment of death as natural comes the ability to fortify yourself psychically and to embrace a philosophical disposition that comforts you.

WHY WE MOURN

When people we love die, whole universes die with them. Those of us still here are not sad for them; we're sad for us. Those people were integral to our existence. Their lives were lamps that lit ours. We loved and were loved by them; suddenly we feel love less and feel less loved. Those people were suns we basked in, and we no longer have those rays to warm us. We're missing something that cannot be restored. What is lost is not just the person, but our relationship to that person. We still have our memories, but not the immediate emotional connection. Different people bring out different facets of our characters. Much of what we are is a reflection in others. Descartes missed something when he concluded, "I think, therefore I am." He omitted the social aspect of human existence: "Others think of me, therefore I am." When someone dies, we lose that part of ourselves, as well as the dead person. We feel diminished by the absence of that person.

Hobbes viewed humans as primarily self-regarding, and these feelings of loss confirm that. Our grief is about ourselves, first and foremost. That is not bad. Don't confuse it with simple selfishness, which disregards others' concerns in favor of your own. After death, we don't know what happens to that person. We hold a variety of beliefs to provide answers, but no one knows for sure. So what we need after someone has died is to let go of that person, to comfort ourselves, and to treasure our memories.

The Tao teaches that we come to know things in comparison with their complements; so with death and life. Those who have had close calls—who walk away from a serious accident or survive cancer against all odds—tell us that they appreciate life more because they have stared death in the face. Most of us take life for granted. We are caught up in satisfying immediate desires, fulfilling long-term goals, and daydreaming in between. Even the Declaration of Independence calls for the pursuit of happiness—if not its attainment. As we saw in the previous chapter, an overarching purpose can be a key to a fulfilling life. But it isn't the end of the story. Too much perspective on the big picture obliterates the value of just one day or even one hour of life. Those who have faced the immediate prospect of no more days or hours understand that value with a clarity most others lack.

Luckily for us, that's an attitude we can cultivate without putting our lives on the line. No recklessness necessary. But you do have to stare death in the face. Not in a catastrophic setting—failing brakes, a lab report reading "malignant," Russian roulette—but with careful contemplation. Right now, most of us enjoy that luxury. If you have the choice, don't wait until you're playing Beat the Clock.

FAITH

Every religion provides answers about what death means. So if you follow a particular faith, you are starting with some maps to guide you on your journey. But you are not at the end of the road. Does a belief in the divine, in an eternal reward, in heaven, make a death painless?

No. The thought of your loved one moving on to a better place is mildly comforting at best. You are still here, with no idea of when or whether you will join the person who has died. Even pastoral counselors admit that saying the dead person is "in God's hands now" isn't enough—though like the rest of us, they often don't know what else to say. (I've learned this from those who have sought philosophical counseling in order to enrich what they offer to those they counsel in turn.)

Hobbes wrote that all religions are derived from fear. From Freud onward, many psychologists and psychiatrists agree: people are terrified of death and so invent happily-ever-after stories (religions) to compensate for their ultimate—but also infantile—anxieties. Whether or not you like that formulation, it helps to remember that humans are generally wary of the unknown. Death is the ultimate example of the unknown. It is preservative—healthy—to be worried about it. If you don't know whether the snake in your path is venomous, it is safer to be afraid of it than to pet it. But beyond addressing fear of the unknown, religions also provide hope: hope for something beyond this world. Religion will help you face the unknown, especially if you have strong faith.

Even those who aren't certain of what they believe can take comfort there. Upon realizing that they are in all likelihood past the halfway point in their lives, some people find religion. Or rediscover it. A cynic would say they are just afraid of death and seeking ready-made reassurances. So what's wrong with that? Nothing reassures like reassurance; beliefs can be comforting. Others don't want to worry about what might or might not happen in a "next world." Either way, the dawning realization that you are, in fact, mortal should only redouble your effort to get the most out of life and to focus on a life of virtue and effort. If nothing else, you'll know you have done the best you can. And if there is something more, you should be favorably judged. That may be hedging your bets, but why not?

Most cultures and religions have a set of conventions for coping with death's aftermath, whether a Jewish shiva, an Irish wake, or any other of countless examples. Most celebrate the person's life, assembling family and friends to eat and drink together and share memories. You might even have a good time. There's usually at least one comedian at every such gathering. All this certainly helps, but only temporarily. Eventually everyone goes home, and you're on your own. We need the ceremonies, and the emotional support of others, and perhaps the spiritual guidance. But those things are usually not sufficient to see us through. When the traditional ceremonies are over, we need philosophical contemplation to carry on.

THE EMPTY TEACUP

Another Buddhist parable teaches us to face death with equanimity. A monk kept a teacup by his bed, and every night before he went to sleep he turned it upside down. Each morning he righted it. When a puzzled novice inquired, the monk explained that he was symbolically emptying the cup of life each night to signify his acquiescence in his own mortality. The ritual reminded him that he had done the things he needed to do that day and so was ready should death come for him. Each morning, then, he turned the cup up to accept the gift of a new day. He was taking life one day at a time, acknowledging the wonderful gift of life with each dawn but prepared to relinquish it at the end of each day.

The first step in creating your own philosophical disposition about death, loss, and mourning is to appreciate life. Living in the moment is the best way to do that. You need an awareness of impermanence to keep you on that path. We know the truth of that, but fool ourselves into thinking that lasting a long time is the same as permanence, and so the end always somehow comes as a surprise. We never think it can happen to us. If you've been driving your whole life without an accident, you are more likely to think you're invincible than simply fortunate. But the more good days we have on this planet, the more grateful we should be for such unpredictable beneficence—without expecting to have it always.

Many people use the present to serve the past or the future. They are busy either ruminating over last week or manipulating tomorrow. They are never in the now. History is past; you can't alter it. The future is uncertain; you can't count on it. What you have for sure is the present. Love being alive right now, and you'll minimize regret when your moments run out.

WHAT YOU BELIEVE

The next step in crafting your own disposition is to reconsider your beliefs about life and death. Use your imagination, and ask yourself, "Where was I before I was born? Where will I be after death?" Visit a cemetery for a sobering experience. Look at all those stones, each of them representing someone who was once a living being with cares and ambitions, enemies and friends, fifteen minutes of fame and dozens of bad hair days. Ask yourself, "Where are they now?" Think about how long, or briefly, they lived and whether all those who remembered them are gone too. So what's important to you in this moment? What's significant for you today? What's vital to you now?

I used to get stuck in traffic jams that crept past a cemetery. When the traffic ground to a dead halt, I could gaze out my window, read the tombstones, and contemplate these issues. I have never been so happy to be alive and stuck in traffic. I noticed that the other drivers, too, were unusually grateful in that vicinity: no honking, cursing, or shooting ever took place around there. I guess they were also happy to be alive and stuck in traffic. So where exactly are you stuck right now? It doesn't much matter, you see, as long as you're happy to be alive.

Plato, Pythagoras, Empedocles, and other ancient Greeks believed in the transmigration of souls (i.e., reincarnation), an idea possibly borrowed from the East. They held that a vital part of who we are survives death and returns in other forms, and so they could view death as not so crushingly final. Western thoughts on death were later generally ceded to Christianity—though the early church maintained a doctrine of reincarnation well into the sixth century.

Hinduism, and some schools of Buddhism, hold that the body dies, but not the spirit, as part of the cycle of birth, death, and rebirth that constitutes the path of spiritual progress. Whether or not you accept the reincarnation portion of that cycle, the idea of the interrelatedness of birth, life, and death helps counter our denial of mortality. In Hindu thought, each lifetime gives you the opportunity to gain knowledge and experience to take with you into the next, thereby gradually progressing toward union with the Godhead. Again, you don't have to believe in reincarnation to benefit from seeing life as a chance to obtain knowledge and experience in the service of progress—and beyond that, a chance to enlist progress in the service of enlightenment.

NOBODY DIES

Classical Buddhism teaches that there is no death because there is no self. The personal self or ego—with all its memories, desires, cravings, anxieties, attachments, and agendas—is an intoxicating mirage that blinds us and distracts us from the unperturbed reality of pure consciousness—our "Buddha nature." The thing you conceive of as you is an illusion. Letting that illusion take your phone calls, attend your meetings, and run your life invites a lot of trouble. The question is: who conceives that illusion of you? The mirage is born, and thus dies. That which conceives it is unborn, and thus undying. Buddhist practice involves putting the self in its place—which is nowhere. Of course the ego doesn't like this, so it tries to prevent you from practicing. Its job is to inflame you, not to make you cool. If you have no sense of self, you cannot experience the death of the self. That's what we fear: the end of personal existence. Hume figured out on his own, and so agreed unknowingly with Buddha, that the self is an illusion. There isn't really anybody in there, so no one really perishes.

> "... when I enter most intimately into what I call myself, I always
> stumble on some particular perception or other, of heat or cold, light or
> shade, love or hatred, pain or pleasure. I never can catch myself without
> a perception, and never can observe any thing but the perception."
> —DAVID HUME

Buddhist philosophy is easily misinterpreted by many westerners, who see something fundamentally repugnant in the intended annihilation of the ego. That, once again, is the ego talking. Of course it can't and won't relinquish its power voluntarily. Only people who have suffered enough (or too much) emotionally realize that it is the self that perpetuates their emotional suffering. If you can take to heart what Buddha (and Hume) meant about the illusoriness of the self, and moreover practice putting your self in its place, then you can handle even death with equanimity.

> "You yourself must make the effort. The Buddhas can only point the
> way. Those who have entered the path and become meditative are
> freed from the fetters of suffering. . . . People beset by craving circle
> round and round, like a hare ensnared in a net . . . they undergo
> suffering for a long time."
> —BUDDHA

The *Tibetan Book of the Dead* describes five stages of after-death existence, called *bardos*, and provides advanced yogas (not the elementary hatha yoga you practiced at yoga camp) designed to guide you through the process. The Tibetans are the only ones I know of who claim to have empirical knowledge of continued existence after death and who teach techniques (to their most advanced yogins) for managing the journey from death to rebirth. This is very different from the more common passive belief in reincarnation. The Tibetans teach how to choose the womb of your rebirth. Not only can you "take charge of your life"; you can possibly even take charge of your death.

In many cultures, experiencing death is likened to solving the essential riddle or mystery. In some Native American traditions, dying is called "crossing the great water." Death is seen as a process of discovery or an important journey. You may require courage to face it, but it is not necessarily something to be dreaded.

SONIA AND ISABELLE, PART II

In Chapter 8, on family, we discussed the case of Sonia, who was rebelling against her mother, Isabelle. When she came to see me, the young woman was struggling mightily against her mother's restrictions and demands. But as we worked together and those issues began to resolve, a further complication arose: Isabelle was diagnosed with terminal cancer.

That's a difficult situation regardless of other circumstances. But Sonia and Isabelle were able to mine it for something valuable. Though their relationship had already begun to ease, they used terminal illness as an opportunity and were now spurred on to reconcile fully. They had been at terrible odds, but impending death put their struggles in perspective for both of them. Their specific conflicts no longer seemed so important, and they made their peace. Death can bring out our humanity; personal differences tend to matter less.

For Sonia, her mother's diagnosis meant having to grow up. Her father wasn't coping with the situation at all, hence much responsibility fell to Sonia as her mother weakened. Reversing the caretaker-caregiver roles called on all Sonia's strength. It allowed her to repay her debt for the care she had received from her mother throughout her life and allowed Isabelle to receive from her daughter at last instead of always giving. Sonia matured through living up to her responsibilities, as difficult as they were.

What we can learn from Sonia and Isabelle's situation is that death illuminates human relations, concentrating the mind like nothing else. So often you hear people say, "I never told him how I felt about him, how much he meant to me" or "I never said how much I loved her" that you'd dismiss it as a cliché if you ran across it in a novel. Death may be inevitable, but losing someone under circumstances that could have been otherwise is the most painful way to experience it. Death's inevitability and its unpredictability give us the strongest reasons to maintain good relations or repair damaged ones. You'll feel the loss more cuttingly if there is something you wanted to say or do but never did. There's nothing worse than being left with the desire to rewind and relive, which of course you can never do.

On the other hand, if death severs a healthy relationship, the pain you'll feel strikes especially deep—you're losing a wholly good thing. But you can find some comfort in knowing that the relationship was everything it could have been.

You don't always get advance notice of death. Lightning strikes. Planes crash. Life can be fleeting. Most of us don't know how long we have left, and we're probably better off not knowing. But don't squander what you have. In this sense, Sonia and Isabelle were actually fortunate. They were able to use Isabelle's remaining time for emotional healing, even when physical healing was no longer possible. It may seem odd to apply the word *lucky* to a daughter facing the loss of her mother. But part of taking a philosophical attitude means looking at positive aspects as well as negative ones. Nothing is gained by ignoring them, and appreciating them might ease your passage through grief.

STELLA

We met Stella in Chapter 10, where I held her up as a positive example of midlife change as opposed to midlife crisis. Since her transition had been relatively smooth, the reason she sought counseling was her struggle to face her own mortality. She was healthy and still relatively young, but hitting "the big five-O" motivated her to search for a way to think about death that made sense to her. The religious faith she was raised with, and that she and her husband had followed when their children were young, no longer resonated for her. At this stage of her life, she felt at a loss when it came to contemplating the end of life—her own and others'—but felt it was important to do so.

As you can see, a successful trip through the midlife minefield did not mean Stella had mastered every aspect of her life. But as she had worked through her issues surrounding midlife change, she had done much of the work of the PEACE process. I taught her the explicit steps of the process, and she was now struggling to integrate what she'd learned in facing other life situations for use in facing death. At this essential level, she sought to build on the dispositions that had helped her through earlier rough patches to confront life's ultimate mystery.

Stella was a very controlling person and tended to exert as much influence as possible on others and on her environment. She took the same approach to her aging, doing all she could to maintain her body and not age earlier or more quickly than necessary. Sometimes her lack of trust interfered with her enjoyment of life, but at least in the area of physical health, she struck a balance. She ate well, exercised, got plenty of sleep, and was even starting to investigate stress-reduction techniques. But she realized that no matter how active you are, there comes a time when you will be less active, and that there are always a number of factors in aging and death that lie beyond our control.

She told me baldly, "I figure you're growing up or you're growing old. I finished growing up a long time ago." She was determined, then, to get the most out of life, but also to acquiesce in aging and the inevitability of death. For her, that's where the going got tough. She knew she had to do it but didn't know how.

The only kind of mourning we really understand is for the loss of others. We're not even very good at that, generally, but we are totally out of our depth when it comes to contemplating our own deaths. Thinking about your own death is always abstract: you can experience someone else's death but not your own (unless the Tibetans are right). It is impossible to conceive of yourself as nonexistent because the very fact that you are trying to conceive of it means that you are existing. In any case, humans have more than enough self-regard to be deeply disturbed by personal extinction. And I don't mean to imply that we're narcissistic, though some therapists would have a field day with you if you walked in with the chief complaint that you were unable to bear the thought of no longer gracing the world with your presence. You'd probably be in therapy until the day you died.

It is natural to think more about your own mortality as you age, though some people never do. But when people realize they've already lived more of their lives than they have left to live, fear of death creeps in unless they are ready with a philosophical defense. Some psychologists make the fear of death the basis of all neurosis. I think a certain amount of trepidation is normal and healthy because, as I've mentioned, fear helps keep us out of dangerous situations. But paralyzing fear is completely unnecessary if you can make your peace with the concept of death.

Equanimity in this arena is especially difficult for people who live life wholeheartedly. Going from the vibrant tapestry of life to, potentially, absolute nothingness can seem like too much to take. That's just to give you fair warning, because that kind of love of life is the very thing I urge you to embrace. But I don't intend to leave you with a complete paradox. Full-out living gives you the best position to finally make peace with death, but it will make the process that much more daunting. At death's door (or upon spotting it far away on the horizon), if you have lived a full and meaningful life, you will know you haven't lived in vain. Having led a decent life, having loved and been loved, having tasted what life has to offer, and having mattered to someone somehow is as much as anyone can expect.

"If this is dying, I don't think much of it."
—LYTTON STRACHEY

TO B-SERIES OR NOT TO B-SERIES

As I mentioned, Western philosophy is relatively limited in its offerings on death. It usually tends toward one of two extremes: a reiteration of Judaeo-Christian faith in an afterlife or a skeptical denial of an afterlife combined with a materialistic anticipation of personal oblivion. In other words: all or nothing. But for those of you who are neither theists nor atheists, an alternative comes from an unexpected direction: an ingenious application of John McTaggart's philosophy of time, by my colleague Stanley Chan. McTaggart's work on time is usually taught as part of the philosophy of physics, but Chan has adopted McTaggart's ideas on time to counsel people facing loss. Chan is a social worker in Toronto to whom terminal cancer patients are referred in cases where physicians can do nothing more than palliate pain.

McTaggart argued that there are two ways to conceive of time: the A-series and the B-series. In the A-series, every moment is either past, present, or future. Every past moment used to be a future moment and a present moment. Thus every moment has the property of being past, present, and future—only at different times. But this is problematic since the terms *past, present,* and *future* already embody a conception of time. Now we need to understand how different points "in" time move "through" time. If we assert that past moments are finished, while future moments haven't yet started, then time dissolves into an ever-moving present moment. As Chan points out, this isn't especially helpful for coming to terms with death—it suggests that the present moment is all there is, and you need to be alive to experience it. When you're dead, your clock isn't ticking any more, so there's no time at all for you.

By contrast, McTaggart's B-series takes a relational view of time: it asserts that every moment happens either before or after every other moment. When one thing has happened before another, then the things happen in this order for all time; the order of events cannot be altered by time's passage. Thus all events in a B-series are forever fixed in terms of their relation to other events. As Chan points out, this implies a sense of endurance or permanence, a kind of record that cannot be obliterated by subsequent events. If one's life is viewed in just that way, it becomes a thread in the tapestry of the B-series. Even when your life ends, it can never "unhappen." All events, including those of your life, are somehow preserved in the B-series of time. You have a slice of immortality—and even a thin slice is better than none.

Chan finds this notion very comforting to dying people who do not believe in an afterlife but who cannot easily face the prospect of oblivion. No one lives forever, but one's life is preserved forever in the B-series. Although not everyone is philosophical enough to see how this can make a difference, in Chan's experience with terminally ill patients, it sometimes makes all the difference in the world. People come to better terms with dying when they see that although death ends life, it does not erase a lifetime. And of course you don't need to be at death's door to appreciate that your life is a sequence of events and that even though that sequence may have a final event, the whole isn't obliterated by that event. This is a way of conceptualizing the lasting meaning and impact of your life on the world without relying on belief in a soul that exists outside the physical limits of the body.

Thinking about the B-series helped Stella come to terms with aging too. Her youth may have passed, but the fact that she was young once— and did all the youthful things that any young person could reasonably expect to do—would never change. That allowed her to make the most of the present without obsessing about the past or fearing the future.

JOANNE

Thinking about the B-series also helped Joanne come to terms with the death of her child. The mother of two grown children, Joanne had five grandchildren. She had been married, widowed, remarried, and divorced. She also managed a successful career. Now in her late fifties, she was full of experience and insight. She knew a lot about herself, about others, and about life. Her problem, however, involved death: the death of her child from cancer, many years ago. Joanne carried on with her life but never got over that loss. Death—and particularly the death of one's child—is perhaps the most challenging problem to handle philosophically.

I didn't use the Buddhist parable of the mustard seed with Joanne, because she had been mourning for so many years that it had become a deep habit. While insights from parables can help prevent the formation of habits, other insights are needed to break them. To show how this worked, let's put her case in the context of the PEACE process.

The problem: Joanne's son, Justin, died of cancer at the age of eight. Joanne refused to accept his death. She continued to mourn for decades and never really got over her grief. Every anniversary of his birth or his death was a terrible time for her.

Emotions: Joanne experienced almost continual sadness and anger, which made her private life terrible and detracted from her relations with family and friends, though she was always effective at work. Psychotherapy and prescription drugs did not alleviate her bad feelings.

Analysis: Joanne wanted to be a good mother and had been a good mother to her living children. She seemed to take it very personally that she was unable to save Jason from cancer, even though she did everything that could have been done at the time. She seemed to interpret his death as evidence that she was an inadequate mother, although this was not so. The only way she could mother her dead child was to mourn him; the only way to continue mothering him was to remain in mourning—which she had been doing for thirty years.

Contemplation: By thinking about the B-series, Joanne began to realize that Justin's eight years of life were not erased by his untimely death. She could look back happily at his years of life and appreciate the gift of his life without sinking into an abyss of despair and self-recrimination over his death. In fact, the best way to mother Justin was to remember him at his best and to realize, through the B-series, that death robs no one of their best moments.

Equilibrium: This new disposition gave Joanne a way to break the long-standing habits that were so debilitating for her private and her social lives. She gradually allowed herself to enjoy life. She would never forget Justin but now began to experience his memory as something pleasant.

GREGORY

My Canadian colleague Stephen Hare had a particularly memorable client, who came to see him asking for one good reason not to commit suicide. Gregory had led a colorful life and enjoyed close relationships with his extended family, but over the last few years (he was almost eighty) had experienced a progressive decline in health. He'd finally been forced to give up his one true passion: skiing. His memory and concentration were not up to snuff, and he feared he might have had an undiagnosed stroke on top of the heart problems he'd been struggling with.

His biggest fear was that he would have a heart attack or stroke that would leave him incapacitated but not dead. He didn't want to be a major burden to his partner or his children, and he didn't want to lose the ability to take his own life if he so desired. His doctors told him he was at risk for a major heart attack or stroke, as he feared. But they couldn't quantify the risk or estimate the time-frame and told him it was far from a certain fate.

His deterioration to this point had been clear but slow. His health problems and fears of future complications left him depressed enough to find his doctor's recommendation of prescription antidepressants interesting, though he hadn't yet tried them.

Despite all this, Gregory revealed that his quality of life was currently very high. Besides a strong relationship with his partner, he had a wide social circle, beloved children and grandchildren, and enough money not to have to worry about it. He regularly went hiking and golfing and attended concerts, plays, and exhibitions. His life was very full, and though he regretted not being able to do all that he once did, he enjoyed his current situation thoroughly. Those he loved would be hard hit by his death, he admitted.

His counselor challenged Gregory with the possibility that he was asking the wrong question, or at least asking it prematurely. Given the many positive aspects of his life and the unpredictability of his future health, what good reason was there to commit suicide? Physical decline, no matter how dismaying, isn't enough to negate the intrinsic value of life when that life is full of love and vitality. Even harsher circumstances, or evidence of imminent incapacitation, might not tip the balance toward suicide, though at some point the balance might shift. Gregory agreed. He already basically recognized that the time had not yet come, which is why he hadn't taken action other than to seek philosophical counseling. His loved ones would rather accept increasing caretaking responsibilities, he finally believed, than lose him prematurely.

He decided to quell his fears by creating a living will with provisions for passive voluntary euthanasia, should it come to that. He knew those provisions were controversial enough to be circumvented sometimes but was willing to balance the risk of that against a current life of quality. The task that remained before him was to enlarge his philosophical perspective to include progressive infirmity as a natural part of aging. Facing the decline takes courage, but of a different kind than Gregory—who was accustomed to physical risk-taking—was used to displaying.

> *"Cowards die a thousand times before their deaths;*
> *The valiant never taste of death but once.*
> *Of all the wonders that I yet have heard,*
> *It seems to me most strange that men should fear;*
> *Seeing that death, a necessary end,*
> *Will come when it will come."*
> —WILLIAM SHAKESPEARE

Gregory actually needed to pull back somewhat from his philosophical stance toward death to look at it from a personal and emotional perspective. That's a line you'll have to find yourself. You don't want to become so nonattached that you lose sight of the value of your life, to yourself and to others.

KEEP AN OPEN MIND

Since we have no proved answers about what happens after death, if you are searching for answers you must consider different possibilities and then settle on the one with the most value to you. Keep an open mind, acknowledging that we don't really know what it means to be alive or dead. Personally, I've experienced too much to be satisfied with the idea of death being absolute nothingness. I think it is conceivable that there's something more, but I've accepted that we don't know one way or the other.

We do know that death can be a very painful separation. Whether or not there is something more after death, the dead person lives on in our hearts. People we love, we love to remember. We should remember the good things and forget the bad—although, as Shakespeare knew, it often works the other way around: "The evil that men do lives after them, the good is oft interred with their bones."

The closer you are to a person, the more painful that person's death will be for you. When two people become as one, in a sense they are reduced to half a person each—neither feels a sense of incompleteness as long as the other provides completion, but neither can feel whole in themselves without the other.

Death can create a gaping hole. Though that kind of attachment inherently brings tremendous potential for pain, I'm by no means saying you shouldn't have such closeness. The secret is to love without selfish attachment. People afflicted by lifelong grief at the death of a loved one are afflicted by their attachment, not by their love. It is possible to love someone with your whole heart while that person is alive, and to love the memory of that person after he dies, and to laugh out loud as you recall some funny thing he said or did, and to shed a tear when you feel sad he isn't here. But it is neither desirable nor necessary to become permanently enveloped in a cocoon of grief. If a part of you has died with a loved one, let go of your attachment and you will become whole again. The quality of your love for the departed will actually improve, and you will no longer be debilitated.

If you need help in letting go of the attachment, there are many philosophical theories and practices you can explore. I've suggested a few in this chapter. In my experience, Buddhist theories and practices are the most reliably effective means of overcoming sorrow—they were designed to do just that and have been refined for two and a half millennia. Buddhist philosophy gives you the healthiest disposition in the face of sorrow; its practices help you substitute constructive habits for destructive ones.

Buddhism has come to the West via many differing traditions—find one that suits you. All have a common denominator: Buddha's Four Noble Truths. First, life entails suffering. Second, suffering has identifiable causes. Third, these causes can be removed. Fourth, appropriate practices remove these causes. The first three truths are theoretical, and there are plenty of books that discuss the theory. The fourth truth involves practice, and there are plenty of places where you can learn to practice. It doesn't matter whether you're religious or nonreligious—after all, religious and nonreligious people suffer alike. Buddhism doesn't care what God(s) you worship or refuse to worship; it cares whether you suffer. When you're ready, it can help you move beyond your personal sorrow.

When physical illness causes suffering, if the illness is healed, the suffering ceases. But people who are not physically ill sometimes suffer needlessly, or suffer more than they have to, from unresolved problems arising from everyday issues of living and dying. Needless or excessive suffering is itself a kind of problem, which this chapter—and this book—have shown you how to manage philosophically. Only when you become tired of needless suffering will you take steps to move beyond it. It's up to you.

"Only when one becomes sick of this sickness can one be free from sickness. The Sage is never sick; because he is sick of this sickness, therefore he is not sick."
—LAO TZU

PART · III

BEYOND CLIENT
COUNSELING

14

Practicing Philosophy with Groups and Organizations

"The crowd is untruth."
—Søren Kierkegaard

"Most of us are at some time or other impelled, even if the impulse is brief, to take a hand in solving the problems of society, and most of us know in our hearts that it is our business to leave the world a little better than we found it."
—Cyril Joad

While the focus of this book has been to explore working with philosophy yourself or in one-on-one counseling, philosophical practice has wider applications as well. Philosophical practitioners work with groups as facilitators and with organizations as consultants. Group facilitation can be informal or formal. Informal groups meet regularly in philosophers' cafés for the purpose of public discussion. Formal groups participate in a process called Socratic Dialogue in order to answer specific questions. As to consulting, the corporate philosopher will become a fixture of twenty-first century organizations, with a reserved parking spot in the company lot. This chapter briefly describes these philosophical activities and tells you how your group or organization can benefit from them.

PHILOSOPHER'S CAFÉS

Europe is already dotted with philosopher's cafés, and they are now mushrooming across North America. There are few technical requirements for such informal philosophical gatherings; all that's required is a willing philosopher to set up shop to moderate discussions every week, every month, or every so often. In our high-tech, fast-paced society, the luxuriously slow-motion exploration of the world of ideas is a unique draw. All sorts of people show up at the groups I've led or attended, but a common feature is often a sense of alienation from mass culture and a realization that since there's diminishing market value to thinking for oneself, it's becoming a lost art.

If you are satisfied with tabloid culture alone—talking heads on TV, no-brainer movies, instant books, disposable lives—you have a steady diet of food for thoughtlessness served up to you every day. But if you're looking for something more, you have to look much harder. In our 57-channels-and-nothing-on world, that quest for something more is increasingly leading to informal philosophical discussion groups. The exchange of ideas is a valuable commodity—despite the lack of a listing for it on Wall Street—and it's mostly free. Philosopher's cafés are bringing philosophy back to its original intention of providing food for thought for people in everyday life and encouraging them to lead more examined lives. Socrates practiced philosophy in the marketplace, taking on all comers, willing to discuss anything with anyone, anytime. This is the tradition of the philosopher's café.

I moderate a monthly Philosopher's Forum in a Manhattan bookstore and another at a famous café in Greenwich Village. There are many regulars, who attend every month, but each time there are also many new faces. The people who come are a cross section of New York—and so a cross section of humanity. They are mostly working people and students. Although it's possible to set a topic for a particular session, I usually let the people gathered raise the issues on their minds and let the group take them where they will.

All sorts of topics are discussed, including the big-ticket items like meaning, morality, faith, and justice. I've facilitated forums on how to overcome alienation, what technology means to humanity, and even how to meet people. The topics covered in Part II of this book often come up in groups, just as they do in one-one-one counseling. Some of the regulars have their own pet topics they like to throw out each time, but no matter what we take up, everyone benefits by hearing others' points of view. You can't expect universal agreement in a public forum. But what you do get is equally useful: an opportunity to challenge other people's views, to have your own views challenged, and to learn to reconcile or tolerate opposing views. Whether the challenge ultimately reinforces or subverts your position, your philosophical stance will be the stronger for it.

There is only one ground rule in my discussion groups: civility. By being civil, the group members also practice other virtues at the same time: patience, attentiveness, tolerance. Independent of the topic under discussion, exercising these virtues is a philosophical lesson in and of itself. I also discourage name-dropping—that is, reference to published philosophical works. Philosophical discussion outside the academy is about what you think, and what the others in the group think—not what someone else has made a career of thinking. If the group is discussing justice, its raw materials are the participants' particular experiences of justice or injustice and their more general thoughts on the matter. You don't need a Ph.D. in philosophy to have experiences and to think for yourself. People who simply drop names, or attempt to impress others with their erudition, are missing the point of the forum.

"For the man who thinks for himself becomes acquainted with the authorities for his opinions only after he has acquired them and merely as a confirmation of them, while the book-philosopher starts with his authorities, in that he constructs his opinions by collecting together the opinions of others: his mind then compares with that of the former as an automaton compares with a living man."
—ARTHUR SCHOPENHAUER

The civility rule is crucial when we push hot-button issues. And believe me, we get into dicier questions than you hear about even on the most outrageous television or radio programs. There are no taboos and no censorship in my philosophical forums, as long as the ground rules are observed to help us exercise reason in tandem with impassioned expression. There are no such things as unthinkable thoughts—just try thinking of one thought you can't think! We take on issues of race, sex, justice, religion, liberty, money, drugs, education, and other topics that are becoming difficult if not impossible to examine openly and sincerely in our increasingly politically correct society.

The guiding purpose of these groups is to discuss things that don't otherwise get discussed—whether because of their awkwardness or their complexity or both. This unbridled exchange of ideas is what America is supposed to be about, so thanks to some bookstores and cafés, we've staked out our monthly territory in which we remain dedicated to individual liberty and freedom of expression. So far, the political commissars have left us alone, the thought police have made no arrests, and the hypersensitive ideologues haven't sued us for offending them. Maybe we need to try harder.

Especially on the touchy subjects, it does us good just to hear other people's takes on things. We usually hang around with like-minded people; I'm betting that most of your friends share most of your views. We are always happy to offer our own two cents worth, but we are often more enriched by someone else's two cents. Hearing other perspectives doesn't necessarily change your mind, but it should make you at least think twice on occasion. Open-mindedness needs to be exercised—you'll need it when your current philosophical disposition isn't serving you well anymore. We have to have opinions, but we don't always know if they are right or wrong. If you want to maintain a high-performance philosophy, you have to give it periodic tune-ups and be willing to make changes when that is called for.

If you are intrigued, I hope you'll seek out—or establish—a philosopher's café in your neighborhood. Bring your big issues with you: Are there limits to social tolerance? What is the purpose of education?

What is the best way to raise children? Do the media wield too much power? Is our culture in decline? What are the ramifications of the written tradition being replaced by a visual one? What does it mean to lead a good life? How do we tell the difference between right and wrong? Are there objective ways to judge what is good and what is evil? Do meaning and purpose exist? Is there a God? Is God a male or female? Does it matter? Is morality reducible to biology? Is morality a human invention? What is beauty? What is truth?

If you've been working on your own or one-on-one with someone else, you may have hit on some of these issues, but more likely you've focused on more immediate, personal concerns. Still, the big and small questions often overlap. Tackling broader topics will reinforce your personal philosophy, which will in turn make that philosophy more persuasive and useful in your everyday life. The big questions are still big. What was discussed in Athens 2,500 years ago remains topical today. Being able to discuss these things is part of what it means to be alive and well.

SOCRATIC DIALOGUE

As informal philosophical discussions burgeon in North America, a more formalized method known as Socratic Dialogue is also taking root. Not to be confused with the Socratic method (to which it bears some relation), Socratic Dialogue is an organized way to answer some big questions. Leonard Nelson—a German philosopher with an English name—outlined the process early in the twentieth century. It was gradually refined by German, Dutch, and lately American practitioners.

To clarify the potentially confusing title of this process, I need to explain why Socrates' name is invoked in two different contexts.

Socrates' theory of knowledge, as reported by Plato, is that we all have it, innately. Asked a stumper like "What is justice?" you probably wouldn't be able to provide a clear definition off the bat, but you would most likely be able to come up with some examples of justice from your own experience. But if you can give an example of something, Socrates would point out that you must already know what that something is—implicitly if not explicitly. This is the basis of Nelson's Socratic Dialogue: a reliable process that guides you to define explicitly what you already know implicitly.

Socrates was also famous for probing people, through a series of questions, until he elicited contradictions from them. If you offered Socrates a careless definition of justice, and if he then led you to admit that your definition could give rise to injustice, you would have thus contradicted yourself. In consequence, your definition could not be correct. Technically this is called the Elenchic method, but it is often known as the Socratic method. Note that it reveals only what something isn't, not what it is. At the end of the day, this method will reveal any number of unserviceable definitions of justice (or whatever's on the table) but not one serviceable one.

By contrast, Socratic Dialogue aims directly at what a thing is. It uses personal experience as the basis for finding a universal definition of the thing at hand that is both explicit and accurate. It employs individual doubt and hard-won consensus to allow you to answer questions like "What is liberty?" or "What is integrity?" That's not the kind of thing you can do on your coffee break; in practice, most Socratic Dialogues are conducted over a full weekend. It takes about two days to get a result with a small group guided by a trained facilitator. Two days is actually a pretty short time considering what's at stake. I mean, you could easily spend a lifetime never knowing exactly what justice, liberty, or integrity are, even though they may be of crucial importance to you. Investing a weekend to catch a close-up glimpse of one of these elusive but eternal ideas is time well spent, in my view. It's like going on a philosophical safari in the big game park of your mind.

The Process

Socratic Dialogues work best with between five and ten participants. That allows enough variety of personal experience, enough time for everyone to participate, and the very real possibility of reaching a consensus. With too few people, there aren't enough points of view to enrich the process. With too many, you'll never get everyone on the same page.

The first step in a Socratic Dialogue is to decide on the question to be answered. Usually that is done beforehand, yet even that part of the process can be an extended educational undertaking. The best questions take the form of "What is X?" with X being liberty, integrity, happiness, fulfillment, hope, love, or any other major but ineffable idea. Other formats can also work, but you can't go wrong with the classic formulation.

Step two is for each participant to think of an example from his or her own life experience that embodies X. It should be a simple example that is no longer ongoing and not too emotional to relate objectively and—if need be—in great detail. Everyone briefly presents his or her example to the group.

Next, the group chooses, by consensus, one example to consider in depth. This will be the primary vehicle for arriving at a definition, but you will be able to find an answer no matter which example you choose. Just choose an example everyone can identify with on some level, to maximize everyone's insight. The selected scenario is then retold in much more detail, and the group poses any clarifying questions they have. No hypothetical questions are allowed. At this stage, and for most of the whole process, it's strictly "Just the facts, Ma'am."

Together the group then breaks down the whole story into its smallest component parts. Even something that happened over a real time span of only a minute or two can have several dozen steps to it. Somewhere in the ordered details, then, is exactly what you seek. It may be in one particular step, or between steps, or in more than one step, or in a combination of steps. Pinpointing the location of X brings you to the halfway mark, because once everyone agrees on where X occurs, you can begin to decide what X is. The whole idea is that if you can capture the actual experience of a thing, you can identify the thing itself. This will be clearer once you consider the example that follows.

Next the group formulates a definition—usually just one sentence—that fits the example at hand. The experience you are focusing on provides a good concrete reference point to check your accuracy. Once you are satisfied, you go back to the other personal experiences and see if they fit the definition you've derived, and modify the definition accordingly. Voilà! A universal definition has been articulated.

The final stage is to try to refute the definition with counterexamples outside of those already presented. This is the only point in the Socratic Dialogue where hypothetical situations are allowed. If you can contradict the definition, you refine it accordingly. You might be surprised at how well your considerately crafted definition stands up in even in this freewheeling phase.

What Is Hope?

A specific example will make the process easier to understand. A group I led recently selected "What is hope?" as the topic for our weekend-long session. To speed us on our way, everyone came to Saturday morning's session having already settled on their personal examples, so we began by presenting them.

One woman talked about how she had hoped to sign a lease on an apartment she and her partner had fallen in love with on first sight. Their lease on their current apartment was expiring, and this new one seemed perfect, but the paperwork kept them in suspense for a few days not knowing whether they would get it (they did). A man discussed the hope he had felt in waiting for a letter from a woman with whom he'd had a brief but intense romance, who had recently moved out of town. She said she would write when she was settled, and he eagerly checked his mail each day, but the letter never came. A journalist said she had hoped a profile she'd written of one of her personal heroes would be picked up by the magazine she thought was the best place for it—which it eventually was. A man doing a lot of volunteer work in his community said he had hoped it would lead to other—paying—jobs. He eventually launched his own business that way. And finally, a woman who had immigrated to America—which required considerable personal and professional sacrifice—discussed the hope she had that her daughter would be able to reap benefits and opportunities unavailable in her home country.

With these choices in front of them, the group elected to work from the experience of the man waiting for a letter. Sam retold his story in detail and answered many questions. Together, the group broke the story down into twenty-three steps (1. In high school, I met two sisters. 2. We all became friends. 3. I took one sister to the prom. . . . 6. Years later, the other sister unexpectedly dropped by my house. . . . 21. I checked the mailbox looking for her letter) and looked for where, exactly, hope came into it. In this case, they found hope in five separate steps, including step 21, above, and step 11: "We made plans to spend time together that coming summer."

Now that they had found hope, we turned to the question "What is hope in this example?" After much deliberation, the group arrived at this definition: "Hope is acting on an expectation for a preferred outcome consistent with one's current life direction." Thus an attempted universal definition emerged from a particular example.

Next we returned to the four other examples presented and modified our definition so that it applied to them as well as to the selected one. Taking the additional complications into account, the group agreed on "Hope is maintaining an expectation for a preferred outcome, consistent with one's current life experience" (substituting *maintaining* for *acting on* and *experience* for *direction*).

This, they were satisfied, applied at every place in every example save one: the woman with hope for her daughter's opportunities didn't think it applied to her. We agreed that hoping for yourself and hoping for someone else are quite different things but found that in the second case, a slight change to our definition would suffice: ". . . consistent with the other's current life experience."

Next we challenged that definition with hypothetical examples, but it withstood our testing. One hypothetical example we considered was Cinderella hoping for Prince Charming to notice her. The debate centered around whether that was within her current life experience and so whether it was hope—as opposed to fantasy—and if so, was it covered by our definition? We allowed that it was, as the scenario was not impossible to her. Perhaps if she herself hadn't thought it possible that a scullery maid would draw the favor of a prince, her hope would have crossed over some border into fantasy. By this time our weekend was nearly at an end, but as far as we got, our definition stood firm.

Many side issues we couldn't resolve in the moment came up during our Dialogue, as they will during any successful Dialogue. The entire process makes you reflect deeply on your experiences, and so it is natural for additional doors to open. In this case, we had to be careful not to get sidetracked by interesting questions like "Does understanding hope require a particular outcome, for better or worse?" "Can you understand hope without knowing what ultimately occurs?" "Are knowing the odds of something happening important to identifying hope?" "Can hope last indefinitely, or is it limited in time?" "Can hope for another reflect altruism or does it always entail self-interest?" "What is the difference between hope and fantasy?" By convention we tabled these issues until the end of the main Dialogue and agreed to take them up then if time allowed. (Despite our hope, we ran out of time.)

It's instructive to compare our consensual definition of hope, arrived at by a group of ordinary people—a writer, a psychologist, an educator, a graduate student, a manager—with definitions offered by some famous philosophers. Hobbes, for example, wrote, "For Appetite with an opinion of attaining, is called Hope." Schopenhauer, for another example, wrote, "Hope is the confusion of the desire for a thing with its probability." Recall, our group wrote, "Hope is maintaining an expectation for a preferred outcome, consistent with one's current life experience." I think our group did as well as if not better than Hobbes, and did much better than Schopenhauer, who obviously got waylaid by a side issue our group had identified: "Are knowing the odds of something happening important to identifying hope?" That a group of thoughtful but otherwise ordinary people can formulate a world-class definition of hope during a single weekend is a testament both to the philosophical understanding that lies dormant in the human mind and to the power of Nelson's method for awakening it.

Socratic Dialogue is an encounter with living wisdom I think everyone should experience at least once. You won't enjoy it until you are personally ready for the undertaking, but there's no better way to get at some of the most complex questions underpinning your life. The trend in America is just beginning (in Germany you can sign up for weeklong Dialogues at resorts). I envision a time when every college student will devote just one weekend—out of four years of classroom lectures—to a Socratic Dialogue. Dialogues can also be facilitated wherever you have willing groups of people: community centers, retirement homes, schools, spas, human development centers. One of my colleagues—Bernard Roy—is marketing them to a cruise line. Sun, sea, and Socrates. Sign me up.

It is the facilitator's job to conduct the group through each of the Dialogue's phases, and in particular to make sure that genuine consensus is reached at each stage before proceeding to the next. Doubts that are not addressed and resolved always crop up in later stages, and by then the group may have taken a wrong turn. The facilitator functions like the conductor of an orchestra, having no voice in the overall score but making sure that everyone plays properly, beautifully, and in unison.

The participants in a Socratic Dialogue also discover something virtually unknown in the West; namely, decision-making by consensus. We have inherited many other models for answering questions, but most of them are deeply flawed. For example, the chain of command hands down directives that must be obeyed whether they are reasonable or not. Or the committee—the method favored by academia—decides all kinds of important issues based on anything but reasonable criteria (e.g., because it's lunchtime) or arrives at a compromise decision nobody agrees with in order to avoid a decision with which some people strongly disagree. Another example: the major weakness of ballot-box elections is that the voters decide on a candidate by making an X rather than by exploring what it means to be the best candidate for the job. Many vote by habit or under the influence of a smear campaign.

But in a Socratic Dialogue, the group gets to the bottom of the issue: right down to its essence. Decisions are methodically weighed and balanced. There is consensus but no compromise. You get the unvarnished truth, or you run out of time trying.

THE CORPORATE PHILOSOPHER

The heyday of American manufacturing, known as the Golden Age of Capitalism, unfolded roughly between the end of World War II and the beginning of the phony energy crisis of 1973. Children born during that prosperous era are called baby-boomers; many of us came of age during the 1960s. That period of unprecedented prosperity in America also witnessed the emergence of an unforeseen profession that contributed to the abundance of the age. Behavioral psychology wedded the manufacturing industry and produced a hybrid offspring: the industrial psychologist. While the Industrial Revolution concentrated on developing ever more efficient machinery and assembly lines, it often ruthlessly exploited and abused its human resources—as Charles Dickens and Karl Marx strongly reminded us. Enter the twentieth century industrial psychologist, who answers these questions: Given state-of-the-art manufacturing processes, how do we produce state-of-the-art employees? We can build efficient machines and design productive assembly lines, but how do we cost-effectively motivate laborers and managers to their utmost efficacy?

The short answer turned out to be something like "Paint the walls green and pipe in Muzak." And even those who dislike green walls and loathe Muzak concede that the industrial psychologist did achieve symbiosis between the muscle of industry and the science of motivation.

In retrospect, he was a primitive precursor of the corporate philosopher. Owing to multinationalism and global civilization, the American economy is shifting from a base of manufactured goods to one of provided services. Formerly the vital linkage was between human bodies and solid machines, and the operational question was "How do you best mechanize human performance?" Answer: "Hire an industrial psychologist and he'll tell you how." Now the vital linkage has changed: it's between human minds and fluid, often amorphous structures, like cyberspace. So the operational question becomes "How do you best systematize human performance?" Answer: "Hire a philosophical consultant and she'll tell you how." That's the general picture.

It's serious, real, and ongoing. In North America, Britain, Europe, and elsewhere, philosophers are working as consultants to governments, industries, and professions—and in all kinds of capacities too. Dilemma training, integrity training, and short Socratic Dialogues are being marketed to government agencies, corporations, and professional organizations.

Some philosophical consultants—like my colleague Kenneth Kipnis—specialize in building mission statements and codes of ethics for organizations, and then design workshops to implement them. You can't just fax (or e-mail) a code of ethics to your workforce and expect them to apply it automatically. Management consultants tried that for years: it never worked. (Then again, it also helps if you know something about ethics, which most management consultants don't.) Employees need to take part in concrete ethical exercises to understand the application of abstract ethical principles and also to anticipate and resolve potential conflicts between their private moralities and their professional codes of conduct. Philosophical consultants provide those services and more.

In America, ethics compliance is a growing area of concern for large corporations, as they are increasingly being held legally liable for actions of individual employees. There are specific federal sentencing guidelines that courts use in awarding damages. If your organization is ethically compliant (i.e., has implemented training in workplace ethics), the financial damages can be significantly decreased. If it's not compliant, they can be significantly increased. The next question is "Whom should you hire to evaluate, design, implement, and follow up your ethics compliance program?" The choice is between a management consultant, who knows next to nothing about ethics but recognizes a business opportunity when he sees one, or a philosophical consultant, who knows more than you'll ever need to know about ethics and who can evaluate, design, implement, and follow up the program you need. It's a no-brainer.

Although cynics might call ethics compliance cheap insurance, I'd call it a strong incentive for a better workplace. It seems pretty clear that virtuous organizations are more functional than vicious ones. Sure you can run a profitable business with the ethics of a snake: all kinds of criminals, scam artists, and occasionally lawyers operate that way. Then again, such people always have to watch their backs, fend off police and other investigators, be prepared for reprisals, and never know when or how their misdeeds will catch up with them. That's neither a good nor an enviable life. You can also run a profitable business with a state-of-the-art ethical agenda, bask in the goodwill of your clients, enjoy the harmony of your workplace, and earn the positive public image that being ethical confers. That's a much more enviable life. Ethics are good for you and good for business. A philosophical consultant builds you a ladder that lets you climb out of the snakepit.

Philosophical practice at the organizational level actually incorporates everything we've discussed in this book. The full-time corporate philosopher counsels individual employees to resolve problems that interfere with getting their jobs done, facilitates workshops with teams of service providers or managers to enhance their performance, and consults with the highest levels of management to improve corporate ethics and dynamics.

Enlightened CEOs might wonder about their responsibility to provide health care to their employees, for example, or about how to approach layoffs or downsizing in the most humane way possible. Employees can consult a philosopher to resolve conflicts among themselves—philosophical intervention in the office might have helped Vincent and his colleague who objected to the painting decorating his office, for example. Or a general topic on workers' minds might be a suitable subject for a philosophical presentation. I conduct workshops for groups of female executives concerned about the "glass ceiling" and how to break through it. I conduct "insensitivity training" workshops for employees who can no longer distinguish the difference between offense and harm. I conduct workshops on ethical integrity and moral worthiness, which help to alleviate or preempt conflicts arising from increasing diversity in the workforce, both ethnicity and gender based.

These philosophical activities are all crucial to getting the most out of work and life with the least amount of friction. As the new millennium dawns, we're way past the point where green wallpaper, Muzak, group therapy, and tranquilizers will make your day. Since Americans are becoming more philosophically minded, America's corporations need to keep in step. In fact, they just might lead the way.

THE LAST WORD

How freely we live depends both on our political system and on our vigilance in defending its liberties. How long we live depends both on our genes and on the quality of our health care. How well we live— that is, how thoughtfully, how nobly, how virtuously, how joyously, how lovingly—depends both on our philosophy and on the way we apply it to all else. The examined life is a better life, and it's within your reach. Try Plato, not Prozac!

PART · IV

ADDITIONAL RESOURCES

APPENDIX A

Hit Parade of Philosophers

This is a brief overview of the sixty-odd philosophers and classic works mentioned in this book whose ideas I find useful in philosophical counseling. There are many others whom I have not mentioned this time around. But herein you will find some of the more important ones.

ARISTOTLE, 384–322 B.C.E.
Greek philosopher, scientist, and naturalist
Themes: logic, metaphysics, ethics
Refrain: the Golden Mean (avoiding extremes in ideals and behavior)
Greatest Hits: *Metaphysics, Nicomachean Ethics*

As a student at Plato's Academy, Aristotle's main concern was knowledge, gathered through observing natural phenomena. He loved to categorize things (he even wrote a book called *Categories*). He virtually invented logic and pioneered several sciences. He also tutored Alexander the Great. For almost two millennia, Aristotle was known as "The Philosopher."

AUGUSTINE, 354–430
North African philosopher and theologian
Theme: original sin
Refrain: Redemption is not in this world.
Greatest Hits: *Confessions, City of God*

Augustine, Bishop of Hippo and a Platonist, happened to be in Rome when it was sacked by Alaric in 410. But Rome had already converted to Christianity and so was supposedly under God's protection. Augustine reconciled this problem by inventing the doctrine of original sin. He is also famous for a prayer in his *Confessions*: "Make me chaste . . . but not yet."

MARCUS AURELIUS, 121–180
Roman emperor and stoic philosopher
Theme: stoicism
Refrain: Do not overvalue what others can take from you.
Greatest Hit: *Meditations*

"Even in a palace it is possible to live well." Marcus Aurelius was not an entirely happy emperor but consoled himself with Stoic philosophy. When people talk about "taking things philosophically," they usually mean stoically—that is, with indifference to worldly pains and pleasures.

FRANCIS BACON, 1561–1626
British philosopher and politician
Theme: empiricism
Refrain: Knowledge is power.
Greatest Hits: *Novum Organum, The Advancement of Learning*

The Godfather of the scientific revolution, Bacon advocated generalizing from specific instances of observed phenomena to scientific laws or theories that could be tested by experiment. He died as a casualty of one of his experiments, developing pneumonia after trying to freeze chickens on Hampstead Heath.

SIMONE DE BEAUVOIR, 1908–1986
French philosopher and feminist
Themes: existentialism, feminism
Refrains: moral responsibility, natural differences between the sexes
Greatest Hits: *The Second Sex, The Ethics of Ambiguity*

Simone de Beauvoir was a stalwart supporter of Jean-Paul Sartre's brand of existentialism, as well as his soul mate. She also wrote eloquently and philosophically about human sex differences and their social consequences.

JEREMY BENTHAM, 1748–1832
British philosopher
Theme: utilitarianism
Refrain: the greatest happiness of the greatest number
Greatest Hit: *Introduction to the Principles of Morals and Legislation*

The founder of utilitarianism, Bentham's primary argument was that actions are moral if they maximize pleasure and minimize pain for those affected by the actions. This is called the "hedonistic calculus." Bentham's waxed bones are clothed and on display in the cloisters of University College London, which he founded. According to his will, his remains are carried into the Senate each year, where he is recorded as "present but not voting."

HENRI BERGSON, 1859–1941
French philosopher and humanist, 1927 Nobel Prize for Literature
Themes: vitalism, dynamism
Refrain: *élan vital* ("life force" not explainable by science)
Greatest Hit: *Creative Evolution*

Bergson criticized mechanistic and materialistic ways of looking at the world, arguing for a more spiritual (but not necessarily religious) approach to life.

GEORGE BERKELEY, 1685–1753
Irish philosopher and bishop
Theme: idealism
Refrain: To be is to be perceived.
Greatest Hits: *A Treatise Concerning the Principles of Human Knowledge, Three Dialogues between Hylas and Philonous*

Berkeley denied the independent existence of material things, arguing that reality is made up of minds and their ideas. Things exist only insofar as they are perceived. Thus Berkeley came close to Buddha's tenet that phenomena are a creation of mind.

BHAGAVAD GITA, 250 B.C.E.–C.E. 250
Ancient Indian epic poem, sixth book of the Mahabharata, author anonymous (attributed to mythical sage Vyasa)
Themes: spiritual consciousness, extinction of unwholesome craving, duty, karma
Refrain: Atman equals Brahma: your personal soul is part of the divine Oversoul.

The *Bhagavad Gita* is full of useful teachings on human suffering, its causes and cures. It espouses the classical doctrine of reincarnation and progress on a spiritual path toward cosmic consciousness.

ANICIUS BOETHIUS, CIRCA 480–524
Roman philosopher, theologian, and consul
Themes: Platonism, Christianity, paganism
Refrain: the use of philosophy to gain perspective on all things
Greatest Hit: *The Consolation of Philosophy*

Boethius, a Roman aristocrat, rose to considerable power before falling from favor and being sentenced to death. He wrote his masterpiece while awaiting execution in prison, and it remains an enduring and inspiring work.

MARTIN BUBER, 1878–1965
German-Jewish philosopher and theologian
Themes: human and human-divine relations
Refrain: I-It versus I-Thou
Greatest Hit: *I and Thou*

To Buber, relationships are either reciprocal and mutual connections between equals or a subject-object relationship involving a degree of control of one over another. Relationships between humans or between humans and God should be of the first order (I-Thou as opposed to I-It).

BUDDHA (SIDDHARTHA GAUTAMA), 563–483 B.C.E.
Indian sage and teacher
Theme: Buddhism
Refrain: how to get beyond sorrow
Greatest Hits: *The Four Noble Truths*, *Dhammapada*, and many *Sutras* (teachings) recorded by his students and followers

Buddha is a title meaning "the enlightened one" or "one who has awakened to the truth." Siddhartha Gautama was the founder of Buddhism. His teachings and practices, which comprise an unorthodox branch of Indian theology/philosophy, show the clearest way to lead a meaningful, useful, compassionate, and pain-free life without invoking religious superstition. Then again, some people practice Buddhism as a religion. Either way, its heart is pure.

ALBERT CAMUS, 1913–1960
French novelist and philosopher, 1957 Nobel Prize for Literature
Theme: existentialism
Refrain: Do the right thing even if the universe is cruel or meaningless.
Greatest Hits: *The Stranger*, *The Plague*

Camus's novels and essays explore the experience of believing in nothing beyond one's individual freedom and actions, and the moral implications of that way of thinking.

THOMAS CARLYLE, 1795–1881
Scottish man of letters, historian, and social critic
Themes: individualism, romanticism
Refrain: Accomplishment is individual.
Greatest Hits: *On Heroes, Hero-Worship and the Heroic in History*

A lapsed Calvinist, Carlyle rejected both mechanistic and utilitarian ways of looking at the world in favor of a dynamic outlook. He believed in the individual morality of a "strong just man" as opposed to the will of the masses and the influence of ordinary events. Interestingly, he also believed that no deceiver could ever found a great religion.

CHUANG TZU, 369–286 B.C.E.
Chinese philosopher-sage, second only to Lao Tzu as renowned Taoist
Theme: Taoism (understanding "the Way," the natural order of things)
Refrain: Learn to attain by *wu-wei* (actionless action).
Greatest Hit: *The Complete Works of Chuang Tzu*

Chuang Tzu was an exemplary Taoist who would not have called himself a Taoist at all. He sought ways to lead a life of benevolence and righteousness, full of humor, free from strife, unbound by social and civil conventions.

CONFUCIUS (KUNG FU TZU), 551–479 B.C.E.
Chinese philosopher, teacher, and government official
Theme: Confucianism
Refrain: Follow the Way through ritual, service, and duty.
Greatest Hit: *Analects*

Confucius advocated government by virtue rather than force. In his view, happiness is achieved by pursuing excellence in personal as well as public life. He upheld piety, respect, religious ritual, and righteousness as the components of harmonious living. His influence on Chinese culture is comparable to Aristotle's influence in the West and may have been greater.

RENÉ DESCARTES, 1596–1650
French philosopher and mathematician
Themes: skepticism, dualism
Refrain: "I think, therefore I am."
Greatest Hits: *Meditations, Discourse on Method*

A founder of modern philosophy, Descartes gave us the full-blown distinction between mind and matter (Cartesian dualism). He emphasized the importance of certainty, achieved through doubt, as the basis of knowledge. He strove to unify the sciences into one system of knowing. He tutored Catherine, Queen of Sweden.

JOHN DEWEY, 1859–1952
American philosopher, educator, and social reformer
Theme: pragmatism
Refrain: Inquiry is self-correcting.
Greatest Hits: *Reconstruction in Philosophy, Experience and Nature, The Quest for Certainty*

Dewey popularized pragmatic, scientific, and democratic ideals. He sought to make educators value the process of inquiry in contrast to the rote transmission of knowledge. Dewey's philosophy was taken to a tragic extreme in latter twentieth century American education, resulting in the demonization of knowledge and the rote transmission of barbarism.

ECCLESIASTES, CIRCA THIRD CENTURY B.C.E.
A King in Jerusalem (Hebrew Koheleth), sometimes identified with Solomon
Theme: life's purpose and conduct
Refrain: "All is vanity, and a striving after wind."
Greatest Hit: Ecclesiastes (a book of the Old Testament)

Ecclesiastes was concerned with the egoism and mortality of man. His writings can be interpreted both optimistically and pessimistically and were sometimes banned by rabbis who thought them too hedonistic. Ecclesiastes has provided titles to novelists (e.g., *Earth Abides* and *The Sun Also Rises*). He gave The Byrds the lyrics to their hit "Turn, Turn, Turn." He also provided several great aphorisms (e.g., "There is nothing new under the sun" and "Cast your bread upon the waters").

EPICTETUS, CIRCA 55–135
Roman philosopher and teacher
Theme: stoicism
Refrain: attachment only to things completely within your own power (such as virtue)
Greatest Hits: *Discourses, Enchiridion*

A freed slave who tutored Marcus Aurelius, Epictetus focused on humility, philanthropy, self-control, and independence of mind. He was said to be more serene than the emperor he served.

EPICURUS, 341–270 B.C.E.
Greek philosopher and teacher
Theme: practical wisdom
Refrain: superiority of contemplative over hedonistic pleasures
Greatest Hits: *On Nature* (fragments survive), *De Rerum Natura* (poem by Lucretius reflecting Epicurean philosophy)

Although Epicureanism has somehow become misidentified with hedonism ("Eat, drink, and be merry, for tomorrow we die"), Epicurus actually advocated moderate pleasures such as friendship and aesthetic pursuits. He founded one of the first communes (The Garden) and regarded philosophy as a practical guide to life. He may have been the original hippie.

KHALIL GIBRAN, 1883–1931
Lebanese-American poet and philosopher
Theme: Arab romanticism
Refrains: imagination, emotion, power of nature
Greatest Hit: *The Prophet*

Gibran's beautiful book of philosophical musings and aphorisms has become a perennial favorite among young adults.

KURT GÖDEL, 1906–1978
Czech-German-American mathematician, logician, and philosopher
Theme: incompleteness theorems
Refrain: Not everything can be proved or disproved.
Greatest Hit: *On Formally Undecidable Propositions of Principia Mathematica and Related Systems I*

Kurt Gödel was able to prove, in 1931, that not every mathematical or logical question can be answered. This effectively put an end to the rationalist quest for perfect and complete knowledge. After emigrating to America, Gödel kept Einstein company at Princeton's Institute for Advanced Study and proved that time travel is not impossible. On the eve of becoming a U.S. citizen, Gödel found a logical flaw in the Constitution that would enable a dictator to take over legally. His friend Oskar Morgenstern convinced him not to bring this to the judge's attention at his swearing-in ceremony.

THOMAS GREEN, 1836–1882
British philosopher
Theme: idealism
Refrain: Being real means being related to other things.
Greatest Hits: introduction to his edition of Hume's work, *Prolegomena to Ethics*

Opposed to empiricism, Green regarded the mind as more than a repository of perceptions, emotions, and experiences but rather as the seat of rational consciousness and capable of producing relations, intentions, and actions. His idea that all our beliefs are interdependent anticipated Quine's famous "web of belief."

GEORG WILHELM FRIEDRICH HEGEL, 1770–1831
German philosopher
Themes: history, politics, logic
Refrain: freedom as self-consciousness in a rationally organized community
Greatest Hits: *The Phenomenology of Spirit*, *The Logic of Hegel*, *Encyclopedia of the Philosophical Sciences in Outline*, *The Philosophy of Right*

Hegel was and remains a very influential philosopher, with wide-ranging ideas about freedom, historical progress, the instability of self-consciousness and its dependence on recognition by others. Unfortunately, Hegel also influenced Marx and Engels and became an unwitting apologist for totalitarian doctrines.

HERACLITUS OF EPHESUS, DIED AFTER 480 B.C.E.
Greek philosopher
Theme: change
Refrain: All things are in a state of flux; you can't step in the same river twice.
Greatest Hit: *On the Universe* (fragments survive)

Heraclitus advocated the unity of opposites and was a proponent of *logos* (reason or knowledge) as an organizing force in the world.

HILLEL, CIRCA 70 B.C.E.–C.E. 10
Babylonian-born rabbi, scholar, and legalist
Themes: morality, piety, humility
Refrain: "What is hateful to you, do not to your neighbor."
Greatest Hit: *Seven Rules of Hillel* (practical applications of Jewish laws)

Hillel was one of the organizers of the first part of the Talmud and an advocate of the liberal interpretation of scripture. He was revered as a great sage, and his students defined Judaism for many generations.

THOMAS HOBBES, 1588–1679
British philosopher
Themes: materialism, authoritarianism
Refrain: Humans are naturally in a war of "all against all" and need a common power "to keep them all in awe."
Greatest Hit: *Leviathan*

Thomas Hobbes founded the fields of political science and empirical psychology. He was the greatest philosopher since Aristotle, and knew it. He wanted for his epitaph "Here lies the true philosopher's stone." His view of humans as supremely egoistic, wildly passionate, easily misguided, constantly power hungry, and therefore highly dangerous beings was enormously unpopular but apparently sound. He argued that politics must not be a branch of theology and that only strong government can prevent violence and anarchy. He made much sense and many enemies. His philosophy anticipated Freudian psychology and provoked the romantic countermovement championed by Rousseau. He tutored Prince Charles II in geometry while in exile during the English Civil War but was forbidden to impart political instruction.

DAVID HUME, 1711–1776
Scottish philosopher
Theme: empiricism
Refrain: "All our ideas are copied from our impressions."
Greatest Hit: *A Treatise of Human Nature*

The outstanding skeptical empiricist, Hume was nicknamed "the infidel." Opposed to Plato, he believed that no ideas are innate. He also denied the reality of the self, the necessity of cause and effect, and the derivation of values from facts. All this made him quite unpopular for a time. He also suggested that metaphysical works be burned and consoled himself with long walks, drinking, and gambling.

I CHING (*BOOK OF CHANGES*), CIRCA TWELFTH CENTURY B.C.E.
Ancient Chinese book of wisdom, author(s) anonymous
Themes: Tao, practical wisdom
Refrain: how to choose wise over foolish courses of action

The *I Ching* maintains that personal, familial, social, and political situations change according to natural laws that the wise understand and take into account when making decisions. By acting in accordance with Tao, one does the right thing at the right time and thus makes the best of any situation. I have been consulting the *I Ching* for thirty years and have never once regretted it.

WILLIAM JAMES, 1842–1910
American psychologist and philosopher
Theme: pragmatism
Refrain: "cash-value" (an idea should be judged on how productive it is)
Greatest Hits: *Principles of Psychology, The Varieties of Religious Experience*

James revealed his dual interests in philosophy and psychology by taking a practical approach to philosophy (pragmatism), believing that an idea is "true" if it has useful results. He emphasized both experimental, laboratory approaches to psychology and analytical reflection on experience.

CYRIL JOAD, 1891–1953
British philosopher and psychologist
Themes: holism, humanism
Refrain: The universe is richer, more mysterious, and yet more orderly than we imagine.
Greatest Hits: *Guide to Modern Thought, Journey through the War Mind*

Joad is a sadly neglected philosopher who believed in enriching the understanding via multiple and equivalently rewarding modes of inquiry: logical, mathematical, and scientific but also aesthetic, ethical, and spiritual. A great moralist and humanist, he was also concerned with the philosophy and psychology of human conflict.

CARL JUNG, 1875–1961
Swiss psychoanalyst and philosopher
Themes: collective unconscious, synchronicity
Refrain: developmental journey toward a final (spiritual) goal
Greatest Hits: *Psychological Types, Synchronicity*

Jung was originally Freud's leading disciple and heir apparent but parted company with him over a major philosophical issue. While Freud postulated a biological basis for every neurosis or psychosis, Jung came to believe that psychological problems are manifestations of unresolved spiritual crises. Jung wrote important introductions to the *I Ching* (Wilhelm-Baynes edition) and the *Tibetan Book of the Dead* (Evans-Wentz edition), making these great works more accessible to the West.

IMMANUEL KANT, 1724–1804
German philosopher
Themes: critical philosophy, moral theory
Refrain: the categorical imperative ("Act only on that maxim which you can at the same time will to become a universal law")
Greatest Hits: *Critique of Pure Reason, Prolegomena to the Metaphysics of Morals*

Kant was a very influential rationalist who tried to ascertain the limits of reason. His theory of morality as duty to higher principles, not anticipation of consequences, is compelling to secular idealists.

SØREN KIERKEGAARD, 1813–1855
Danish philosopher and theologian
Theme: existentialism
Refrains: free will, individual choice
Greatest Hits: *Either/Or, The Sickness unto Death*

Kierkegaard—the first existentialist—rejected Hegel's systematic philosophy as well as organized religion. In his view, human judgment is incomplete, subjective, and limited, but we are also free to choose and responsible for our choices. Only by exploring and coming to terms with fundamental anxieties can we become liberated within our ignorance.

ALFRED KORZYBSKI, 1879–1950
Polish-American philosopher
Theme: general semantics
Refrains: Humans are uniquely aware of time ("time-binding" animals). Conventional socialization and language promote unnecessary conflicts.
Greatest Hits: *Science and Sanity, Manhood of Humanity*

Korzybski is another neglected but important philosopher, who viewed the human animal as in its collective childhood and suggested ways in which we might eventually mature as a species. He explained how structures of language and habits of thought condition and trigger destructive emotions and sought ways to restructure our thinking.

LAO TZU, CIRCA SIXTH CENTURY B.C.E.
Chinese philosopher
Theme: Taoism
Refrains: complementarity of opposites, attainment without contention, harmonious relations
Greatest Hit: *Tao Te Ching* (*The Way and Its Power*)

Lao Tzu's identity and the century he lived in are still debated, but regardless, his ideas on living a life in harmony with the Way remain powerful and influential. He appears to have been a senior civil servant who wrote down his philosophy upon retirement—apocryphally at the behest of a border guard who wouldn't let him leave the province otherwise. He penned a truly great philosophical guide, and thereby founded Taoism.

GOTTFRIED WILHELM LEIBNIZ, 1646–1716
German mathematician, philosopher, and historian
Theme: rationalism
Refrain: This is the best of all possible worlds.
Greatest Hits: *New Essays Concerning Human Understanding, Theodicy, Monadologie*

While Voltaire lampooned Leibniz's belief that this is the "best of all possible worlds" via the character Dr. Pangloss in *Candide,* Leibniz believed that everything happens for sufficient reasons, many of which we cannot understand. Leibniz (at the same time as Newton) co-invented the calculus; he also invented binary numbers. He believed in free will.

JOHN LOCKE, 1632–1704
British philosopher and physician
Themes: empiricism, science, politics
Refrains: Experience is the basis of knowledge; the human mind is a tabula rasa (blank slate) at birth.

Greatest Hits: *Essay Concerning Human Understanding, Two Treatises of Government*

Locke is one of the important British empiricists. Originally a physician, he saved the life of the Earl of Shaftesbury by innovatively inserting a pipe to drain an abdominal abscess. This led him into favor with powerful people, who sought his philosophical advice. Politically, Locke argued for individual liberties and constitutional rule, which placed him ahead of his time in England and exerted considerable influence on nascent American political thought.

NICCOLO MACHIAVELLI, 1469–1527
Italian consiglieri
Theme: political philosophy
Refrain: To be a successful leader you must act in whatever way works, without concern for conventional morality.
Greatest Hit: *The Prince*

With realism that was shocking at the time, Machiavelli declared that the world is not a moral place and that politics, particularly, is not an ethical enterprise. Bertrand Russell called *The Prince* "a handbook for gangsters," but I'd say it's more like *Despotism for Dummies.*

JOHN MCTAGGART, 1866–1925
British philosopher
Theme: idealism
Refrain: Reality is more than material.
Greatest Hit: *The Nature of Existence*

McTaggart believed that there was no God but did believe in individual immortality. His philosophy of time (B-series) provides an enduring account of endurance.

JOHN STUART MILL, 1806–1873
Scottish philosopher, economist, and politician
Themes: utilitarianism, libertarianism, egalitarianism
Refrain: individual liberty
Greatest Hits: *On Liberty, Utilitarianism, A System of Logic, On the Subjection of Women*

Mill thought restrictions on individual liberty should be allowed only to prevent harm to others and was an ardent advocate of free speech, individual responsibility, and social egalitarianism. His brand of utilitarianism differed from Bentham's in that Mill thought pleasure was not the sole measure of good. "Better Socrates dissatisfied than a pig satisfied," he asserted.

GEORGE EDWARD MOORE, 1873–1958
British philosopher
Themes: analytic philosophy, idealism
Refrains: "the defense of common sense"; goodness cannot be defined but is intuitively understood
Greatest Hit: *Principia Ethica*

Moore is most famous for his so-called naturalistic fallacy, the mistake he claims we make when we try to identify good with any naturally existing object or property or try to measure it in any way. Nonetheless, Moore asserted that actions can be right or wrong, even though goodness cannot be defined.

IRIS MURDOCH, 1919–1999
British philosopher and novelist
Themes: religion and morality

Refrain: the reinstatement of purpose and goodness in a fragmented world
Greatest Hit: *The Sovereignty of Good*

Murdoch revived Platonism as an antidote to the lack of meaning and morality in the twentieth century world. She conveyed her philosophy primary and artfully through novels.

LEONARD NELSON, 1882–1927
German philosopher
Theme: synthesis of rationalism and empiricism
Refrain: We can reason from our particular experiences to arrive at an understanding of universals.
Greatest Hit: *Socratic Method and Critical Philosophy*

Nelson made an invaluable contribution to philosophical practice by developing the theory and method of Socratic Dialogue. When properly applied, Nelsonian Socratic Dialogue provides definitive answers to universal questions such as "What is liberty?" "What is integrity?" and "What is love?"

JOHN VON NEUMANN, 1903–1957
Hungarian-American mathematician and philosopher
Themes: game theory, computing, physics
Refrain: Decision-making in situations of risk, conflict of interest, or uncertainty can be analyzed through game theory to find the best choice.
Greatest Hit: *Theory of Games and Economic Behavior* (with Oskar Morgenstern)

John von Neumann contributed brilliantly to several fields, including mathematics, computing theory, and quantum mechanics. His invention (with Morgenstern) of the theory of games marked the inception of an entirely new branch of mathematics, which has applications for philosophy, psychology, sociology, biology, economics, and political science—not to mention philosophical counseling.

FRIEDRICH NIETZSCHE, 1844–1900
German philosopher
Theme: extravagant anticonventionalism
Refrains: the will to power, man versus superman
Greatest Hits: *Thus Spake Zarathustra, Beyond Good and Evil, The Genealogy of Morals*

Philosopher, poet, prophet, and syphilitic, Nietzsche is rarely dull. He despised mainstream society and castigated Christianity as a religion for slaves. He advocated the emergence of an *übermensch* (superman), who would transcend conventional morality—an idea badly abused by the Nazis. Interestingly, he also appeals to postmodernists, whose politics tend toward the other extreme. This is a testament to Nietzsche's genius (or possibly lunacy). He penned pithy aphorisms and cooked up much provocative food for thought (e.g., "God is dead," "Socrates was rabble").

CHARLES SANDERS PEIRCE, 1839–1914
American philosopher and scientist
Theme: pragmatism
Refrain: Truth is an opinion we all ultimately agree on and represents an objective reality.
Greatest Hit: *Collected Papers*

Peirce was the founding figure of American pragmatism, which was further and differently developed by Dewey and James. To distinguish his version from James's, Peirce coined the term *pragmaticism,* which didn't quite catch on. Peirce's philosophy was criticized by Russell for its apparent subjectivity, but in fact Peirce was very scientific in his outlook.

PLATO, CIRCA 429–347 B.C.E.
Greek philosopher and academician
Theme: essentialism
Refrain: The essences of goodness, beauty, and justice can be understood only through a philosophical journey.
Greatest Hit: *The Dialogues of Plato* (including *The Republic*)

Plato founded the Academy (the prototypical university) in Athens. His dialogues involving his teacher Socrates comprise most of what we know about Socrates' philosophy, so the ideas of Plato and Socrates can be difficult to tease apart. Plato is considered to be the founder of philosophical study and discourse as still practiced today.

PROTAGORAS OF ABDERA, CIRCA 485–420 B.C.E.
Greek philosopher and teacher
Themes: relativism, sophism
Refrain: "Man is the measure of all things."
Greatest Hits: known primarily through Plato's dialogues "Protogoras" and "Thaetetus"

Protagoras believed that moral doctrines can be improved upon, even if their value is relative. He also believed that virtue can be taught. He developed dialectical and rhetorical methods later popularized by Plato as the Socratic method. *Sophistry* has acquired an undeservedly pejorative connotation. The sophists taught people, for a fee, how to argue persuasively for any given point of view, no matter how patently false or unjust. Thus the sophists trained the first generation of lawyers.

PYTHAGORAS, BORN CIRCA 570 B.C.E.
Greek philosopher and mathematician
Themes: metempsychosis and mathematics
Refrain: All things are based on geometric forms.
Greatest Hits: Pythagorean theorem, Pythagorean comma

More is attributed to Pythagoras than is known about him. He apparently taught the doctrine of metempsychosis (the transmigration of souls, or reincarnation), and refrained from eating beans. He is credited with the famous theorem of Euclidean geometry named after him. He is also credited with the discovery that the twelve-tone (diatonic) musical scale does not permit instruments to be tuned perfectly. This anomaly eventually led to equal temperament tuning at the time of J. S. Bach (e.g., *The Well-Tempered Clavier*).

WILLARD QUINE, 1908–
American philosopher
Theme: analytical philosophy
Refrain: All beliefs depend on other beliefs.
Greatest Hit: *From a Logical Point of View*

Quine is the most important American philosopher of the latter half of the twentieth century. His contributions began in logic and set theory and continued in theories of knowledge and meaning. He is famous for challenging Kant, for shifting away from logical positivism, and for reframing Green's idea that beliefs are always held in conjunction with other beliefs.

AYN RAND, 1905–1982
Russian-born American writer and philosopher
Themes: objectivist ethics, romantic capitalism (libertarianism)
Refrains: the virtues of egoism, the vices of altruism
Greatest Hits: *The Fountainhead, Atlas Shrugged, The Virtue of Selfishness*

Ayn Rand is an important original thinker who championed integrity and ability as keys to a productive and prosperous society. In her view, capitalism without exploitation (enlightened self-interest) is the best system; socialism with exploitation (unenlightened collective interest) is the worst. Rand's fictional capitalists are all schooled in philosophy, and they are all virtuous beings.

WILLIAM ROSS, 1877–1971
British philosopher
Theme: theory of prima facie obligations
Refrain: Some duties must be more stringently followed than others; the priority depends on each case.
Greatest Hits: *The Right and the Good, Foundations of Ethics*

Ross points out that duties come into conflict, in the sense that we often have to fulfill one obligation at the expense of another. His theory suggests that we must prioritize our duties carefully, according to each situation.

JEAN-JACQUES ROUSSEAU, 1712–1778
Swiss philosopher
Theme: romanticism
Refrain: The human being is born a "noble savage" and is corrupted by civilization.
Greatest Hits: *The Social Contract, Discourse on the Origin and Bases of Inequality among Men*

Rousseau focused on the rift between man and nature and the tension between intellect and emotion, recommending nature and emotion as the higher ways of being. Although his romanticism provides a counterbalance to Hobbes's authoritarianism, Rousseau's philosophy of education is a recipe for disaster.

BERTRAND RUSSELL, 1872–1970
British philosopher, 1950 Nobel Prize for Literature
Themes: realism, empiricism, logic, social and political philosophy
Refrain: "Philosophy is an unusually ingenious attempt to think fallaciously."
Greatest Hits: *Principia Mathematica* (with Alfred North Whitehead), *History of Western Philosophy, Human Knowledge: Its Scope and Limits, Unpopular Essays*

Russell published more than seventy books in his lifetime; his philosophical analyses ranged over every conceivable subject. He was a great and learned man who did not shy away from political causes and social controversy. He was famously denied a position at The City College of New York after a New York State court declared him to be an immoral influence on society, mostly because of his then avant-garde (now common-place) views on open marriage and divorce. While the Athenians killed Socrates for allegedly corrupting its youth, Americans merely denied Russell employment. Russell might have conceded that this implied social progress.

JEAN-PAUL SARTRE, 1905–1980
French philosopher and novelist, 1964 Nobel Prize for Literature
Themes: existentialism, politics, Marxism
Refrain: free will; "bad faith" (denying responsibility for our actions)
Greatest Hits: *Nausea, Being and Nothingness, Existentialism Is a Humanism*

Sartre was the leading French intellectual of his age. He studied with Husserl (the founder of phenomenology) and Heidegger (the leading German figure in existential-ism). A Marxist by conviction, Sartre attempted to found a political party in France. Notwithstanding his Marxist commitments, he staunchly defended his belief in indi-vidual responsibility.

ARTHUR SCHOPENHAUER, 1788–1860
German philosopher
Themes: volition, resignation, pessimism
Refrain: The will stands outside space and time, but following its dictates leads to misery in no time at all.
Greatest Hit: *The World as Will and Idea*

Schopenhauer was well educated, fluent in many European and classical tongues, and had a notoriously difficult relationship with his mother. He is famous for trying and failing to dislodge Hegel, whom he regarded as a sophist and charlatan, from his position of influence. He sought refuge from emotional suffering in Indian philosophy. He wrote pungent essays and acerbic aphorisms and was one of the few philosophers whom Wittgenstein read or admired. Whether this bodes well or ill for Schopenhauer depends on whether you read or admire Wittgenstein.

LUCIUS SENECA, 4 B.C.E.–C.E. 65
Roman philosopher and statesman
Themes: stoicism, ethics
Refrain: Philosophy, like life, should be primarily about virtue.
Greatest Hit: *Moral Letters*

Seneca rose from obscurity in provincial Cordoba to become Emperor Nero's tutor, lieutenant, and ultimately his victim. Seneca lived and died according to the moral dictates of Stoicism, enduring hardship, triumph, and death with equanimity. He committed suicide in the Roman tradition, by opening his veins in a hot bath, when commanded to do so by the insanely paranoid Nero.

SOCRATES, CIRCA 470–399 B.C.E.
Greek philosopher and teacher
Theme: the Socratic method
Refrain: The good life is the examined life, spent pursuing wisdom at all costs.
Greatest Hits: Socrates' ideas are preserved only in Plato's dialogues, so it is sometimes difficult to separate Socrates the man from Socrates the character and to distinguish between Socrates' thoughts and Plato's.

The historical Socrates and the historical Plato are easier to separate. Socrates (like Buddha, Jesus, and Gandhi) was an influential sage who had no official employment or position but whose wisdom attracted important followers and who has grown in stature since his death. Socrates saw himself as a political gadfly, constantly stinging Athenians into awareness of their philosophical shortcomings. He allowed himself to be put to death by the corrupted state because his reasoned argument compelled him to remain even though his friends had arranged his escape. Thus he prized philosophy above life itself. Plato never forgave the Athenians for executing Socrates. Christians believe that Jesus died to redeem mankind from sin; it may be secularly asserted that Socrates died to redeem philosophers from unemployment.

BARUCH SPINOZA, 1632–1677
Dutch philosopher and lens grinder
Theme: rationalism
Refrain: All knowledge can be deduced.
Greatest Hits: *Tractatus Theologico-Politicus, Ethics*

Spinoza's views managed to get him expelled from the Jewish community, and his writings were also attacked and banned by Christian theologians. He even attracted hostility in tolerant Holland, where he had taken philosophical refuge. Spinoza believed that self-preservative human passions (i.e., appetites and aversions) lead to predetermined acts but that we can become free by liberating our reason from the shackles of passion. Like Hobbes, Spinoza thought that we do not like something because it is good; rather, we call something good because we like it.

SUN TZU, CIRCA FOURTH CENTURY B.C.E.
Chinese military adviser
Theme: philosophy of warfare
Refrain: "Being unconquerable lies with yourself."
Greatest Hit: *The Art of War*

Sun Tzu redefined conflict as a philosophical art form. He taught that the "pinnacle of excellence" is to subjugate your foe without fighting. His philosophy of warfare can be applied analogously to many other kinds of human conflict, from marital strife to office politics.

HENRY DAVID THOREAU, 1817–1862
American writer, poet, and philosopher
Theme: New England transcendentalism (libertarianism)
Refrain: the "unquestionable ability of man to elevate his life by conscious endeavor"
Greatest Hits: *Walden, On Civil Disobedience*

Thoreau advocated simplicity, individual responsibility, and communing with the natural environment as keys to the good life. He lived and breathed his philosophy. His theory of civil disobedience exerted seminal influence on both Gandhi and Martin Luther King.

ALFRED NORTH WHITEHEAD, 1861–1947
British philosopher
Theme: empiricism
Refrain: Natural science should study the content of our perceptions.
Greatest Hits: *Principia Mathematica* (with Bertrand Russell), *The Concept of Nature, Process and Reality*

Whitehead sought a unified interpretation of everything from physics to psychology.

LUDWIG WITTGENSTEIN, 1889–1951
Austrian philosopher
Theme: philosophy of language
Refrains: the scope and limits of language; language as a social instrument
Greatest Hits: *Tractatus Logico-Philosophicus, Philosophical Investigations*

Wittgenstein believed that philosophy has at least one "therapeutic" task: to clarify misunderstandings and imprecisions of language, which themselves give rise to philosophical problems. He is one of the most influential philosophers of the twentieth century.

MARY WOLLSTONECRAFT, 1759–1797
British philosopher and feminist
Theme: egalitarianism
Refrain: Social function should not be based on sex difference.
Greatest Hits: *Vindication of the Rights of Women, Vindication of the Rights of Men*

Wollstonecraft was ahead of her time in asserting women's rights. She wrote articulately and persuasively in favor of egalitarianism. Her correspondence with the great conservative Edmund Burke is illuminating. She was also the mother of Mary Shelley, who wrote *Frankenstein*.

Appendix B

Organizations for Philosophical Practice

AMERICAN ORGANIZATIONS

AMERICAN PHILOSOPHICAL PRACTITIONERS ASSOCIATION (APPA)

The APPA is a national professional association founded in 1998. It trains, certifies, and represents philosophical practitioners across the spectrum of client counseling, group facilitation, and organizational consulting. It also fosters professional and educational relations with other counseling professions, with organizations, and with the public. The APPA offers the following categories of membership:

• *Certified memberships* are offered to experienced philosophical practitioners—client counselors, group facilitators, or organizational consultants—who meet APPA requirements. Certified members are listed in the *APPA Directory of Certified Practitioners* and are eligible for referrals and other professional benefits. Certified members are bound by the APPA Code of Ethical Professional Practice.

• *Affiliate memberships* are offered to recognized counseling or consulting professionals in other fields (e.g., medicine, psychiatry, psychology, social work, management, law) who wish to be identified with and become better acquainted with philosophical practice but who do not seek APPA certification. Affiliate members are eligible to attend special events, meetings, and workshops.

• *Adjunct memberships* are offered to holders of an accredited M.A. or Ph.D. in philosophy or to those with equivalent philosophical backgrounds who wish to become certified members. Adjunct members are eligible to attend APPA certification training programs, from Primary (Level I) to Advanced (Level II), completion of which enables them to become certified members. Adjunct members are also eligible to attend special events, meetings, and workshops.

• *Auxiliary memberships* are offered to friends and supporters of philosophical practice. The APPA Auxiliary welcomes all who wish to join in this capacity. No qualifications are necessary beyond an interest in leading a more examined life. Auxiliary members are eligible to attend special events, meetings, and workshops.

• *Organizational memberships* are offered to all entities (e.g., corporations, institutions, professional associations) seeking to benefit or become beneficiaries of philosophical practice. Organizational members are eligible for a range of philosophical services provided by certified members of the APPA, including ethics compliance, ethos enhancement, employee workshops, and executive seminars.

All APPA members receive the newsletter, invitations to events, and other benefits.

The APPA is a not-for-profit educational corporation. It admits certified, affiliate, and adjunct members solely on the basis of their respective qualifications and admits auxiliary and organizational members solely on the basis of their interest in and support of philosophical practice. The APPA does not discriminate with respect to either members or clients on the basis of nationality, race, ethnicity, sex, gender, age, religious belief, political persuasion, or other professionally or philosophically irrelevant criteria.

Membership forms and other information can be obtained by e-mail, by fax, from the APPA website, or by writing to the APPA. Please address all inquiries to:

APPA
The City College of New York
137th Street at Convent Avenue
New York, NY 10031
tel: 212–650–7827
fax: 212–650–7409
e-mail: info@appa.edu
http://www.appa.edu

President: Lou Marinoff
Vice Presidents: Vaughana Feary, Thomas Magnell, Paul Sharkey
Secretary-Treasurer: Keith Burkum

AMERICAN SOCIETY FOR PHILOSOPHY, COUNSELING AND PSYCHOTHERAPY (ASPCP)
The ASPCP (founded 1992) is an open academic society dedicated to exploring the relation between philosophy, counseling, and psychotherapy. It holds annual meetings in conjunction with the American Philosophical Association (APA).

Please address inquiries to:

Vaughana Feary
President Elect, ASPCP
37 Parker Drive
Morris Plains, NJ 07950
tel/fax: 973–984–6692

FOREIGN NATIONAL ORGANIZATIONS

CANADA

Canadian Society for Philosophical Practice

473 Besserer Street
Ottawa, Ontario K1N 6C2
Canada
Stephen Hare, Interim President
tel: 613–241–6717
fax: 613–241–9767

GERMANY

International Society for Philosophical Practice (formerly German Society for Philosophical Practice)

Hermann-Loens Strasse 56c
D–51469 Bergisch Gladbach
Germany
Gerd Achenbach, President
tel: 2202–951995
fax: 2202–951997
e-mail:
Achenbach.PhilosophischePraxis@t-online.de

ISRAEL

Israel Society for Philosophical Inquiry

Horkania 23, Apt. 2
Jerusalem 93305
Israel
tel: 972–2–679–5090
e-mail: msshstar@pluto.mscc.huji.ac.il
http://www.geocities.com/Athens/Forum
/5914
Shlomit Schuster, Chief Inquirer

NETHERLANDS

Dutch Society for Philosophical Practice

Wim van der Vlist, Secretary
E. Schilderinkstraat 80
7002 JH Doetinchem
Netherlands
tel: 33–314–334704
e-mail: W.vanderVlist@inter.nl.net
Jos Delnoy, President
Herenstraat 52
2313 AL Leiden
Netherlands
tel: 33–71–5140964
fax 33–71–5122819
e-mail: ledice@worldonline.nl

NORWAY

Norwegian Society for Philosophical Practice

Cappelens vei 19c
1162 Oslo
Norway
tel: 47–88–00–96–69
e-mail: filosofiskpraksis@bigfoot.com
http://home.c2i.net/aholt/e-nsfp.htm
Henning Herrestad, President
e-mail: herrestad@online.no
Anders Holt, Secretary
e-mail: aholt@c2i.net
tel: 47–22–46–14–18
cell: 47–92–86–43–47

SLOVAKIA

Slovak Society for Philosophical Practice

Dept. of Social & Biological
Communication
Slovak Academy of Sciences
Klemensova 19, 81364 Bratislava
Slovakia
Emil Visnovsky, President
tel: 00421–7–375683
fax: 00421–7–373442
e-mail: ksbkemvi@savba.sk

UNITED KINGDOM

Anglo-American Society for Philosophical Practice (AASPP)

Keynes House, Austenway
Gerrards Cross, Bucks SL9 8NW
United Kingdom
Anne Noble, Secretary
tel: 01753–981874
fax: 01753–889419

Co-Chair, UK: Ernesto Spinelli
Co-Chair, US: Lou Marinoff

Society of Consultant Philosophers

The Old Vicarage
258 Amersham Road
Hazlemere nr High Wycombe
Bucks HP15 7P2
United Kingdom
tel: 01494 521691
e-mail: 106513.3025@compuserve.com

Chair: Karin Murris
Secretary: Elizabeth Aylward

APPENDIX C

Directory of Philosophical Practitioners

Note: Most philosophical practitioners identified themselves for listing in this directory; others were listed as a courtesy. Philosophical practice is currently unregulated in America and worldwide. Individual methods, styles, and orientations may vary considerably.

As this book goes to press, the American Philosophical Practitioners Association (APPA) has begun to certify qualified practitioners. Consult the APPA or its website (see Appendix B) for an up-to-date directory of APPA-certified counselors, facilitators, and consultants.

AMERICA

Wills Borman
6752 Dogtown Road
Coulterville, CA 95311
tel/fax: 209–878–0304
e-mail: wborman@usa.net
counselor, facilitator, consultant

Jon Borowicz
Milwaukee School of Engineering
1025 North Broadway
Milwaukee, WI 53202–3109
e-mail: BOROWICZ@warp.msoe.edu
counselor

Keith Burkum
5620 Moshulu Avenue, Apt. 1
Bronx, NY 10471
tel/fax: 718–543–3885
tel: 973–778–1190, ext. 6140
e-mail: Speedo6034@aol.com
counselor, consultant

Harriet Chamberlain
1534 Scenic Avenue
Berkeley, CA 94708
tel: 510–548–9284
fax: 510–540–1057
counselor, facilitator, consultant

Elliot Cohen
Department of Philosophy
Indian River Community College
3209 Virginia Avenue
Fort Pierce, FL 34981–5599
e-mail: COHENE@mail.firn.edu
clinical philosophy

Kenneth Cust
Department of English & Philosophy
Central Missouri State University
Warrensburg, MO 64093
tel: 816–543–8775
e-mail: kencust@sprintmail.com
counselor

Barbara Cutney
782 West End Avenue, Apt. 81
New York, NY 10025
tel: 212–865–3828
counselor

Richard Dance
6632 East Palm Lane
Scottsdale, AZ 85257
tel: 602–945–6525
fax: 602–429–0737
e-mail: rdance@swlink.net
counselor

Peter Dlugos
Bergen Community College
400 Paramus Road
Paramus, NJ 07652
tel: 201–447–9282
e-mail: pdlugos@mailhost.bergen.cc.nj.us
counselor

Mary van Eepoel
4210 W. Fig Street
Tampa, FL 33609
tel: 813–288–2059
counselor

Kathy and James Elliott
Anthetics Institute
3108 Pinehook Road, Suite 101
Lafayette, LA 70508
counselors

Vaughana Feary
37 Parker Drive
Morris Plains, NJ 07950
tel/fax: 973–984–6692
e-mail: VFeary@aol.com
counselor, facilitator (including
Socratic Dialogue), consultant

Claude Gratton
University of Nevada at Las Vegas
4505 Maryland Parkway, Box 455028
Las Vegas, NV 89154
tel: 702–895–4333
fax: 702–895–1279
e-mail: grattonc@nevada.edu
counselor, consultant

Pierre Grimes
Academy for Philosophical Midwifery
5122 Bolsa Avenue, Apt. 102
Huntington Beach, CA 92649
e-mail:
mind@philosophicalmidwifery.com
counselor

Victor Guarino
442 Easy Street
Sebastian, FL 32958
tel: 561–388–0184
counselor

David Hilditch
7152 Tulane Avenue
University City, MO 63130
tel: 314–727–1675
e-mail: Hilditch@excite.com
counselor

Alicia Juarrero
4432 Volta Place, NW
Washington, DC 20007
tel: 202–342–5128
fax: 202–342–5160
e-mail: ja83@umail.umd.edu
counselor

Stephanie and Jonathan Kastin
1115 Prospect Avenue, Apt. 301
Brooklyn, NY 11218
tel: 718–369–0010
e-mail: jkastin@post.harvard.edu
counselors

Christopher Keller
555 Buena Vista West, Apt. 703
San Francisco, CA 94117
tel: 415–863–8017
counselor

Kenneth Kipnis
Department of Philosophy
University of Hawaii at Manoa
2530 Dole Street
Honolulu, HI 96822
e-mail: kkipnis@hawaii.edu
facilitator, consultant

Ran Lahav
RR 3, Box 115
West Glover, VT 05875
tel: 802–525–6275
e-mail: lahavr@construct.haifa.ac.il
counselor

Thomas Magnell
Department of Philosophy
Drew University
Madison, NJ 07940
tel: 973–267–2582
e-mail: tmagnell@drew.edu
consultant

Lou Marinoff
Department of Philosophy
The City College of New York
137th Street at Convent Avenue
New York, NY 10031
tel: 212–650–7647
fax: 212–650–7409
e-mail: marinoff@cnct.com
counselor, facilitator (including Socratic
Dialogue), consultant, video-
conferencing available

Christopher McCullough
1510 Eddy Street, PH3A
San Francisco, CA 94115
tel/fax: 415–885–1037
e-mail: CMccull787@aol.com
counselor

James Morrow
1055 West Morrow Street
Elba, AL 36323
tel: 334–897–6522
counselor

G. Steven Neely
900 Powell Avenue
Cresson, PA 16030
tel: 814–886–4219
counselor

Roger Paden
Department of Philosophy & Religious
Studies
George Mason University
Fairfax, Virginia 22030
tel: 703–993–1290
fax: 703–993–1297
counselor

Benjamin Phillips
188 Mang Avenue
Kenmore, NY 14217
tel: 716–876–4919
e-mail:
Phillips.Benjamin_B@buffalo.va.gov
counselor

Dennis Polis
1145 Linden Avenue, Apt. 3
Glendale, CA 91201
tel: 818–845–6499
counselor

Sidney Rainey
9513 Kentstone Drive
Bethesda, MD 20817
tel: 301–564–0253
fax: 301–564–1296
facilitator, consultant

Ross Reed
3778 Friar Tuck Road
Memphis, TN 38111
tel: 901–458–8112
e-mail: DoctorReed@yahoo.com
counselor

Tj Reilly
6195 Raasaf Circle
Las Cruces, NM 88005
tel: 505–523–1325
e-mail: simone@zianet.com
counselor, consultant

Susan Robbins
957 Broadway
Westville, NJ 08093
tel: 609–742–5500
e-mail: srobbins@netreach.net
counselor

Bernard Roy
396 Third Avenue, Apt. 3N
New York, NY 10016
tel: 212–686–3285
e-mail: bernard_roy@baruch.cuny.edu
counselor, facilitator (including Socratic
Dialogue), consultant

Mehul Shah
Baruch College CUNY
Department of Philosophy, Box G–1437
17 Lexington Avenue
New York, NY 10010
tel: 914–591–7488
fax: 212–387–1728
e-mail: mshah1967@aol.com
facilitator (including Socratic Dialogue)

Paul Sharkey
Box 222
Lancaster, CA 93584
tel: 805–726–0102
e-mail: PWSHARKEY@email.msn.com
counselor, consultant

Wayne Shelton
Center for Medical Ethics
Albany Medical College
47 New Scotland Avenue, Apt. 153
Albany, NY 12208
tel: 518–262–6423
fax: 518–262–6856
e-mail: wshelton@ccgateway.amc.edu
counselor, consultant

Richard Stichler
Department of Philosophy
Alvernia College
Reading, PA 19606
tel: 610–779–6266
fax: 610–370–9163
e-mail: rnstich@aol.com
counselor, facilitator, consultant

AUSTRALIA

Stan van Hooft
Faculty of Arts
School of Social Inquiry
221 Burwood Highway
Burwood, Vic 3125
Australia
tel: 61–3–9244–3973
fax: 61–3–9244–6755
e-mail: stanvh@deakin.edu.au
facilitator

Steven Segal
35 Memorial Avenue
St. Ives 2075
Australia
tel: (Sydney) 9144–4382
counselor

CANADA

Hakam Al-Shawi
Department of Philosophy
York University
4700 Keele Street
Toronto, Ontario M3J 1P3
Canada
tel: 416–422–3987
e-mail: hakam@yorku.ca
website:
www.geocities.com/HotSprings/Villa/
3253/
counselor

Stanley Chan
270 Old Post Road
Waterloo, Ontario N2L 5B9
Canada
tel: 519–884–5384
fax: 519–884–9210
e-mail: stanley_chan@pmh.toronto.on.ca
counselor

Stephen Hare
473 Besserer Street
Ottawa, Ontario K1N 6C2
Canada
tel: 613–241–6717
fax: 613–241–9767
e-mail: share@ced.flora.org
website: www.ced.flora.org/dr-hare
counselor, facilitator

David Jopling
Department of Philosophy
York University
4700 Keele Street
Toronto, Ontario M3J 1P3
Canada
tel: 416–736–2100 ext.77588
fax: 416–736–5114
email: jopling@yorku.ca
counselor, consultant

Peter March
Department of Philosophy
St. Mary's University
Halifax, Nova Scotia B3H 3C3
e-mail: Peter.March@StMarys.ca
facilitator

Cheryl Nafziger-Leis
16 Meadowlark Road
Elmira, Ontario N3B 1T6
Canada
tel: 519–669–4991
fax: 519–669–5641
e-mail: Leis@sentex.com
counselor, consultant

Justine J. Noel and Susan M. Turner
P.O. Box 5364
Victoria, British Columbia, V8R 6S9
Canada
tel: 250–370–9070
fax: 250–721–8728
e-mail: jnoel@uvic.ca, sturner@uvic.ca
counselors

Peter B. Raabe
46–2560 Whiteley Court
North Vancouver, British Columbia,
V7J 2R5
Canada
tel: 604–986–9446
e-mail: raabe@interchange.ubc.ca
counselor

FINLAND

Antti Mattila, M.D.
Tykistonkatu 11 B 30
SF–00260 Helsinki
Finland
counselor

FRANCE

Anette Prins-Bakker
43, Avenue Lulli
92330 Sceaux
France
tel: 33–014–661–0032
fax: 33–014–661–0031
e-mail: prins@aol.com
counselor, facilitator

GERMANY

There are many practitioners in Germany.
For an up-to-date listing, contact the
International Society in Appendix B.
Here is a partial listing.

Gerd Achenbach
Hermann-Loens Strasse 56c
D–51469 Bergisch Gladbach
Germany
tel: 49–2202–951995
fax: 49–2202–951997
e-mail:
Achenbach.PhilosophischePraxis@t-
online.de
counselor

Christian Hick
Ebernburgweg 9–11
50739 Köln
Germany
tel: 49–0221–9171085
fax: 49–0221–9171086
e-mail: Christian.Hick@Koeln.netsurf.de
counselor

Ute Maria Kraemer
Neue Linner Str. 15
47799 Krefeld
Germany
tel: 49–2151–618365
counselor

Eckart Ruschmann
P.O. Box 608
79006 Freiburg
Germany
fax: 49–761–39728
counselor

Christoph Weismüller
Am Dammsteg 59
40591 Düsseldorf
Germany
tel: 49–0211–228122
counselor

ISRAEL

For an up-to-date listing, consult the
Israel Society in Appendix B. Here is a
partial listing.

Lydia Amir
The College of Management
9 Shoshana Persitz Street
Tel-Aviv, 61480
Israel
tel: 972–3–690–2091
fax: 972–3–699–0460
counselor

Ora Gruengard
43 Yehuda Hanasi Street
Tel-Aviv, 69391
Israel
tel: 972–3–641–4776
fax: 972–3–642–2439
e-mail: egone@mail.shenkar.ac.il
counselor

Ran Lahav
25 Yasmin Street
Mevasseret Zion, 90805
Israel
tel: 972–2–534–3047
e-mail: lahavr@construct.haifa.ac.il
counselor

JAPAN

Narifumi Nakaoka
Department of Philosophy
Faculty of Letters
The University of Osaka
1–5 Machikaneyama-cho, Toyonaka-shi
560 Japan
tel: 06–850–5662
fax: 06–850–5836
e-mail: nana@let.osaka-u.ac.jp
clinical philosophy

LUXEMBOURG

Jean-Luc J. Thill
Conseil et Recherche en Philosophie
10, rue du XI Septembre
L–9282 Diekirch
Luxembourg
tel: 352–80–87–44
fax: 352–80–87–45
e-mail: info@philosophie.lu
website: www.philosophie.lu
counselor

NETHERLANDS

There are many practitioners in the
Netherlands. For an up-to-date listing,
consult the Dutch Society in Appendix
B. Here is a partial listing.

Dries Boele
Spaarndammerplantsoen 108
1013 XT Amsterdam
Netherlands
tel: 31–20–686–7330
counselor, facilitator (Socratic Dialogue),
consultant

Jos Delnoy
Herenstraat 52
2313 AL Leiden
Netherlands
tel: 33–71–5140964
fax 33–71–5122819
e-mail: ledice@worldonline.nl
counselor, facilitator (Socratic Dialogue),
consultant

Ida Jongsma
Hotel de Filosoof
(Philosopher's Hotel)
Anna Vondelstraat 6
1054 GZ Amsterdam
Netherlands
tel: 31–20–683–3013
fax: 31–20–685–3750
facilitator (Socratic Dialogue), consultant

Will Heutz
Schelsberg 308
6413 AJ Heerlen
Netherlands
tel: 31–45–572–0323
counselor

Jos Kessels
Dialogue Consultants
Nieuwe Kerkstraat 147
1018 VL Amsterdam
Netherlands
tel: 31–20–625–7036
fax: 31–20–620–0872
e-mail: dialogue@xs4all.nl
facilitators (Socratic Dialogue),
consultants

Henk Van Luijk
Michelanglostraat 62
1077 CG Amsterdam
Netherlands
tel: 31–20–675–7634
fax: 31–20–679–4709
e-mail: vanluijk@nijenrode.nl
consultant

Eite Veening
Geerten Gossaertlan 10
9721 XJ Groningen
Netherlands
tel: 31–50–272248
counselor

Yvonne Verweij
Uilenburgstraatje 10
5211 ED s-Hertogenbosch
Netherlands
tel/fax: 31–73–613–8126
counselor

Ria Vriend
Vlamingstraat 13
2011 WR Haarlem
Netherlands
tel: 31–23–531–5023
fax: 31–20–641–8970
counselor, facilitator (Socratic Dialogue)

NORWAY

For an up-to-date listing, consult the
Norwegian Society, Appendix B. Here is a
partial listing.

Anders Lindseth
University of Tromso
N–9037 Tromso
Norway
e-mail: andersl@fagmed.uit.no
counselor

SLOVAKIA

Emil Visnovsky
Department of Social & Biological
Communication
Slovak Academy of Sciences
Klemensova 19

813 64 Bratislava
Slovakia
tel: 00421–7–375683
fax: 00421–7–373442
e-mail: ksbkemvi@savba.sk
counselor

Pavol Vyletet
Masarykova 32
984 01 Lucenec
Slovakia
tel/fax: 00421–863–4320688
counselor

SOUTH AFRICA

Barbara Norman
Department of Education
University of Witwatersrand
WITS 2050, Johannesburg
South Africa
e-mail: 022bar@mentor.edcm.wits.ac.za
counselor

SWITZERLAND

Willi Fillinger
Lavaterstrasse 66
8002 Zürich
Switzerland
counselor

TAIWAN

Jess Fleming, Ph.D.
Tamkang Univerity
Tamsui, Taipei County
Taiwan, R.O.C.
e-mail: fleming@hpap.tku.edu.tw
counselor

UNITED KINGDOM

Emmy van Deurzen
Warden's House
33 Shore Lane
Sheffield S10 3AY
United Kingdom
tel/fax: 0114–268–4025
e-mail: Emmy@compuserve.com
counselor

Tim LeBon
7 Ormonde Gate
Chelsea, London SW3 4EU
United Kingdom
tel: 0171–542–8912
or
19 Irwin Road
Guildford, Surrey GU2 5PW
United Kingdom
tel: 0148–345–7893
e-mail: timlebon@aol.com
website:
http://members.aol.com/timlebon
counselor

Simon du Plock
School of Psychotherapy & Counselling
Regent's College
Inner Circle, Regent's Park
London NW1 4NS
United Kingdom
tel/fax: 0171–487–7406
counselor

Ernesto Spinelli
School of Psychotherapy & Counselling
Regent's College
Inner Circle, Regent's Park
London NW1 4NS
United Kingdom
tel/fax: 0171–487–7406
counselor

APPENDIX D

Further Reading

BOOKS

Achenbach, Gerd; *Philosophische Praxis*, Köln: Jürgen Dinter, 1984.

Cohen, Elliot; *Philosophers at Work*, New York: Holt, Rinehart & Winston, 1989.

Deurzen, Emmy van; *Paradox and Passion in Psychotherapy*, New York: John Wiley & Sons, 1998.

Eakman, Beverly; *Cloning of the American Mind: Eradicating Morality through Education*, Lafayette, La.: Huntington House Publishers, 1998.

Ehrenwald, Jan (editor); *The History of Psychotherapy*, Northvale, N.J.: Jason Aronson, 1997.

Erwin, Edward; *Philosophy and Psychotherapy*, London: Sage Publications, 1997.

Evans-Wentz, W. (editor); *Tibetan Yoga and Secret Doctrines*, London: Oxford University Press, 1958.

Grimes, Pierre; *Philosophical Midwifery*, Costa Mesa, Calif.: Hyparxis Press, 1998.

Hadot, Pierre; *Philosophy as a Way of Life*, London: Blackwell, 1995.

Held, Barbara; *Back to Reality: A Critique of Postmodern Theory in Psychotherapy*, New York: W. W. Norton, 1995.

Kapleau, Philip; *The Three Pillars of Zen*, New York: Doubleday, 1969.

Kennedy, Robert; *Zen Spirit, Christian Spirit*, New York: Continuum, 1997.

Kessels, Jos; *Socrates op de Markt, Filosofie in Bedrijf*, Amsterdam: Boom, 1997.

Koestler, Arthur; *The Ghost in the Machine*, London: Hutchinson, 1967.

Lahav, Ran, and Tillmanns, Maria (editors); *Essays on Philosophical Counseling*, Lanham, Md.: University Press of America, 1995.

McCullough, Chris; *Nobody's Victim: Freedom from Therapy and Recovery*, New York: Clarkson Potter, 1995.

Nelson, Leonard; *Socratic Method and Critical Philosophy*, trans. Thomas Brown III, New York: Dover Publications, 1965.

Russell, Bertrand; *A History of Western Philosophy*, New York: Simon & Schuster, 1945.

Spinelli, Ernesto; *The Interpreted World*, London: Sage Publications, 1989.

Szasz, Thomas; *The Myth of Mental Illness*, New York: Harper & Row, 1961.

Thome, Johannes; *Psychotherapeutische Aspekte in der Philosophie Platons*, Hildesheim, Zurich, New York: Olms-Weidmann, (*Altertumswissenschaftliche Texte und Studien*, Vol. 29), 1995.

Wallraaf, Charles; *Philosophical Theory and Psychological Fact*, Tucson: University of Arizona Press, 1961.

Wiseman, Bruce; *Psychiatry: The Ultimate Betrayal*, Los Angeles: Freedom Publishing, 1995.

Woolfolk, Robert; *The Cure of Souls: Science, Values and Psychotherapy*, San Francisco: Jossey-Bass Publishers, 1998.

SCHOLARLY JOURNALS ON PHILOSOPHICAL PRACTICE

Zeitschrift fur Philosophische Praxis (Journal for Philosophical Practice)
> Michael Schefczyk, editor
> Grabengasse 27
> 50679 Köln
> Germany
> (published in German and English)

Filosofische Praktijk (Philosophical Practice)
> Dutch Association for Philosophical Practice (see Appendix B)
> (published in Dutch)

International Journal of Applied Philosophy
> Elliot Cohen, editor
> Philosophy Program
> Indian River Community College
> 3209 Virginia Avenue
> Fort Pierce, FL 33454–9003

Journal of Applied Philosophy
> Society for Applied Philosophy
> Carfax Publishers, Abingdon, Oxfordshire, United Kingdom

Journal of the Society for Existential Analysis
> Hans W. Cohn and Simon du Plock, editors
> Society for Existential Analysis
> BM Existential
> London WC1N 3XX
> United Kingdom

SPECIAL EDITIONS ON PHILOSOPHICAL PRACTICE

Journal of Chinese Philosophy
> Chung-Ying Cheng, editor
> Vol. 23, No. 3, Sept. 1996
> Philosophical counseling and Chinese philosophy

Inquiry: Critical Thinking Across the Disciplines
> Robert Esformes, editor
> Vol. 27, No. 3, Spring 1998
> Selected papers from the Third International Conference on Philosophical Practice

APPENDIX E
Consulting the *I Ching*

"The I Ching insists upon self-knowledge throughout. The method by which this is to be achieved is open to every kind of misuse, and is therefore not for the frivolous-minded and immature . . . It is appropriate only for thoughtful and reflective people who like to think about what they do and what happens to them . . ."
—CARL JUNG

The *I Ching*, or *Book of Changes*, predates Confucius and Lao Tzu. Both of these great philosophers refer to it, and its author (or authors) is unknown. Full of ancient and enduring wisdom, the *I Ching* has been widely misunderstood in the West, and translated for many different purposes. Many use it as a fortune-telling device or oracle. In truth, it is a treasure-trove of philosophical wisdom, and, if used properly, a mirror to what you are actually thinking, even subconsciously. The method of selecting yarrow stalks or tossing coins to direct you to a particular passage is irrelevant to getting a clear reflection of what is in your own heart and mind—what you are perhaps on the verge of knowing consciously, but what has not yet emerged. This Appendix briefly outlines the proper way to use coins to point you to a reading.

It is this interactive component that has got the *I Ching* dismissed (or hailed) as a fortune-teller. The coins will direct you to a reading and, in my experience, the selected passage often does have a remarkable ability to speak directly to a person's immediate issues. Some believe that there is a mystical connection between your tossing of the coins and the answers they lead you to in the book. That need not be the case. If you just opened the book randomly at any page, you would be more likely to open it somewhere nearer the middle than at either end. You'd hit the beginning or ending chapters with less probability than the middle ones. The coins may simply even out the odds. But no matter which chapter you hit, your active conscious mind will find something meaningful and useful in the text, which is actually a reflection of what is meaningful and useful in your submerged thoughts. There will be a resonance between its wisdom and yours, for the *I Ching* mirrors what is in your heart.

You could also read the whole book and pick out what applies to you. But it's hundreds of pages long, so that would take you a while. Anyway, each chapter is dense with insight. One chapter offers more than enough each time you wish to consult the book.

I suggest you try the *I Ching* with the method outlined below. I prefer the Wilhelm-Baynes translation far better than the rest—it's the only one I'd dream of using. But you'll have many editions to choose from in most bookstores.

The *I Ching* is divided into sixty-four thematic chapters, and each chapter is identified with a hexagram—a figure composed of six lines. Each line is either broken or unbroken (dashed or solid). Broken lines are considered yin lines (female), and unbroken are yang (male). That way of categorizing the lines reminds us that they are not opposites, but complements, as embodied in the familiar Taoist symbol. In Chinese philosophy, the balance between them is critical.

There are two ways to select your hexagram—and so which chapter you should consult at a given time. One way involves manipulating yarrow stalks, and is more complicated. But you can use three ordinary coins to arrive at the same result. Tradition dictates using the humblest coin in deference to the exalted advice you seek, so I use pennies (though any coin would work). Take three coins and toss them, and you will end up with one of four possible combinations: three heads, three tails, one head and two tails, or two heads and one tail. By convention, heads has a value of 2; and tails, 3. Add up the value of your combination (3 heads equals 6; two heads and one tail equals 7; two tails and one head equals 8; three tails equals 9). You've translated a coin toss into a number, and now you translate the numbers into a line: 6 or 8 indicates a yin (broken) line; 7 or 9 indicates a yang (unbroken) line. Each yin or yang line can also be "changing" or "unchanging." 6 and 9 are changing lines; 7 and 8 are unchanging lines.

Throw the coins a total of six times, and each time write down the yin (broken) or yang (unbroken) line you obtain on a piece of paper. Also indicate (by an asterisk or other mark) any changing lines. Make sure you start from the bottom up: your first throw is the bottom line of the hexagram; your last throw, the top line. Once you get the hang of it, this whole process will take just a minute or two.

JASON

For example, Jason had been volunteering his time to serve an organization whose mission he believed in. His service was recognized as valuable, and the Board of Directors invited him to assume a position of considerable responsibility—still on a voluntary basis—which Jason accepted. The organization flourished. But not everyone on the Board was Jason's friend. One or two had opposed his appointment, possibly because of envy or other personal problems they were experiencing. One of Jason's opponents— call him George—had Jason investigated, without his knowledge, by an independent expert to whom George had given selected documents about Jason's activities. Based on this limited evidence, the independent expert concluded that Jason was in conflict of interest between his volunteer work for this organization and his other professional endeavors. Jason wanted to maintain both relationships: with the organization and with his profession. But George wanted Jason to resign from the organization, and tried to use this investigation to force his resignation.

Jason was trying to decide whether to defend himself against this unwarranted attack by George, or to counter-attack by asking the Board to get rid of George instead, among other things. So he consulted the *I Ching* (as well as his friends on the Board). These are the lines he obtained: 8,8,7,7,9,9. The corresponding hexagram looks like this:

```
———      *
———      *
———
———
— —
— —
```

The asterisks indicate the changing lines: a nine in the fifth place, and a nine in sixth place.

The hexagram is number 33, whose theme is "Retreat." (You find your hexagram's number and location in the book by looking it up in a table provided in most editions.) Among other things, Retreat advised:

"Flight means saving oneself under any circumstances, whereas retreat is a sign of strength . . . Thus we do not simply abandon the field to the opponent; we make it difficult for him to advance by showing perseverance in single acts of resistance. In this way we prepare, while retreating, for the counter-movement."

Jason understood this to mean that he should not immediately resign from his position, that he should resist the allegation of "conflict of interest," but that he should not counter-attack George (at least, not at this time). Moreover, Retreat offered some important advice about Jason's attitude, as well as his conduct:

"The mountain rises up under heaven, but owing to its nature it finally comes to a stop. Heaven on the other hand retreats upward and remains out of reach. This symbolizes the behavior of the superior man toward a climbing inferior; he retreats into his own thoughts as the inferior man comes forward. He does not hate him, for hatred is a form of subjective involvement by which we are bound to the hated object. The superior man shows strength (heaven) in that he brings the inferior man to a standstill (mountain) by his dignified reserve."

That seemed clear enough: George's allegations would never hold water before the whole Board, so Jason could defeat him without confronting him or hating him. (Similar advice is offered in *Matthew* 5:39—". . . resist not evil".)

When you get changing lines, as Jason did in the fifth and sixth places, you change them to their opposites (yin into yang, yang into yin) and thereby obtain a new hexagram, which is meant to address the situation that follows this one. A nine changes to an eight, while a six changes to a seven. Changing Jason's two nines into eights, his new hexagram looks like this:

```
— —
— —
 —
 —
— —
— —
```

This is hexagram 62, "Preponderance of the Small." It gave Jason further advice on his continued relationship with the organization:

"Exceptional modesty and conscientiousness are sure to be rewarded with success; however, if a man is not to throw himself away, it is important that they should not become empty form and subservience, but be combined always with a correct dignity in personal behavior. We must understand the demands of the time in order to find the necessary offset for its deficiencies and damages."

Thus the *I Ching* helped Jason maintain his relationship with the organization, and fend off the misconduct of others without inviting further retribution on himself.

When consulting the *I Ching*, take your time to ponder the reading you obtain to discover what wisdom it holds for you in your current situation. Once you're satisfied with at least your initial impressions, this is where "changing" and "non-changing" lines come in. This is the *Book of Changes*, after all, so this part is crucial. If you have no changing lines, that means your situation is relatively constant for now. On the other hand, if you have all changing lines, that means you have a lot of stuff going on. Most often, you'll fall somewhere in between (as Jason did). There's no absolute time-frame in any case—the changes might be happening now, or next week, or in five years. You will know what it means for you. You determine the rate of the change. Each chapter has additional special commentary for each changing line.

Any hexagram can potentially change into any other. Symbolically, that gets to the heart of this philosophy: Any situation can change into any other. Practically, this means that the *I Ching* doesn't address just 64 specific situations, but many more than that. Within each hexagram, there are 64 possible combinations of changing and unchanging lines (e.g. no changing lines, all changing lines, and every possibility in between). There are also 64 different hexagrams, which thus account for 64 times 64, or 4,096 possible situations. Moreover, any hexagram with changing lines (that is, any of 4,032 possible changing situations) can change into any other hexagram (64 possible situations). Thus the *I Ching* addresses a total of 4,032 times 64, or 258,048 possible situations. (By contrast, the Daily Horoscope column addresses 12.)

The *I Ching*, like Chinese philosophy generally, is not dependent on fate. It is big on making the best of a given situation. The focus is on your part in the current circumstances—and the wisest way to play it. Your present will become one of many possible futures, but you can engender desirable outcomes and prevent undesirable ones by assuming responsibility and being prudent. What you think, say, and do is up to you. The *I Ching* guides you in going from a bad situation to a good one, or a good one to a better one, and helps you avoid going from a bad one to a worse one. Consider this passage from hexagram 15, "Modesty":

"The destinies of men are subject to immutable laws that must fulfill themselves. But man has it in his power to shape his fate, accordingly as his behavior exposes him to the influence of benevolent or destructive forces."

Whereas much of Western philosophy begins with common sense but arrives at paradox, much of Chinese philosophy begins with paradox but arrives at common sense.

After you throw the coins and look up your reading, you may be amazed at how specifically it applies to whatever is on your mind, as it did in both Sarah's and Jason's cases. I have no explanation as to how this book so often pushes just the right button—but I gladly accept the result. Call me a mystical pragmatist. Jung called it "synchronicity"—a non-coincidental correspondence between two apparently unrelated sequences of events. You seek advice, you throw some coins, and the *I Ching* gives you expert guidance purely by chance. It doesn't exactly add up, but it surely works. Hume thought that "chance" was a vulgar word, expressing only our ignorance. The *I Ching*'s reliability suggests that Hume was right.

You might also be led to a hexagram whose advice is not transparently obvious. In either case, the point is to use what you are reading as a springboard for contemplation. Plato would approve the premise: knowledge is already within you, though you may need some help getting it out.

I see the *I Ching* as a window that gives you a view of what you really think about things. Psychoanalysts and psychologists have their windows, too, for looking into your mind. Freudian techniques of free-association of words (the analyst says a word, then you say the first word that pops into your head), and the interpretation of dreams, are two ways of getting at your deeper thoughts and feelings. This can also be done nonverbally, as with Rorschach (ink-blot) tests. Consulting *I Ching* is a philosophical way of fishing in these deep waters. But instead of finding something wrong with you, it helps you do the right thing. And therein lies a world of difference.

Index